# ENERGY AND EFFORT

SYMPOSIA OF THE
SOCIETY FOR THE STUDY OF HUMAN BIOLOGY

Volume XXII

# ENERGY AND EFFORT

*Edited by*
G. A. HARRISON

TAYLOR & FRANCIS LTD
LONDON
1982

First published 1982 by Taylor & Francis Ltd, 4 John Street,
London WC1N 2ET

© 1982 Taylor & Francis Ltd

Printed and bound in Great Britain by Taylor & Francis (Printers) Ltd.
Rankine Road, Basingstoke, Hampshire RG24 0PR.

**British Library Cataloguing in Publication Data**

Energy and effort.—(Symposia of the Society for
   the Study of Human Biology, ISSN 0081-153X; 22)
   1. Bioenergetics—Congresses
   I. Harrison, G. Ainsworth    II. Series
   574.19'121          QH510

   ISBN 0-85066-224-9

# CONTENTS

vi                          *Contents*

# PREFACE

The papers published in this volume represent the contributions that were delivered at the 49th meeting of the Society
for the Study of Human Biology, which was held at St. John's
College, Oxford on January 7th and 8th, 1981 under the chair-
manships of Professors J.S. Weiner, J.M. Tanner, D.F. Roberts
and the Editor of this volume. Grateful acknowledgement is
made to the Royal Society for a grant which helped finance the
meeting, particularly the participation of the overseas speakers,
and to St. John's College for its many considerations. The
Editor, who was the organizer of the meeting, also wishes to
record his indebtedness to the many members of the staff of
the Department of Biological Anthropology at Oxford who helped
him and particularly to his colleague, Dr. A.J. Boyce.

Why did the Society choose 'Energy and Effort' as a theme
for a symposium meeting? One of the important areas of growth
in human biology, and especially biological anthropology, in recent years, has been in human ecology. A knowledge of the
way human beings affect and are affected by their environment
is not only of great intellectual interest in itself, but is also of
some substantial practical concern to environmental management,
health and population regulation. The academic focus is also
a wide-ranging one, for ecological dynamics represent the evolutionary forces which shaped the past and which will produce the
future. Most of the studies in human ecology, especially those
dealing with environmental effects on man, have been rather restricted ones, concerned with particular problems or hypotheses
in, for example, nutrition, climatic physiology, or epidemiology.
Whatever else they may have shown, they have always demonstrated how difficult it is to isolate the effects of single environmental or human factors, since in the real world everything operates in concert with innumerable interactions. We are only beginning to be able to deal with this, since analysis necessitates

the availability of enormous amounts of data, preferably on exactly the same people and means of handling these in complex processing. Before the arrival of modern computing facilities, it was an impossibility.

Now that we can for the first time begin to examine the functioning of whole human populations in a holistic way, it is necessary to identify foci of attention. The energetics of a population provide one such focus. There is nothing a community does, or does not do, which hasn't an energy implication, and much of what it does is actively concerned with energy acquisiton. But to focus attention solely on energy intakes and expenditures, costs, balances and flows would be to treat the human individual or group only as a machine. People feel the costs of energy expenditure and energy intake; they have, or at least think they have, choice, and they can and do spend a great deal of time doing things, and giving priority to things which, viewed only in terms of energy, are nonsensical. To understand even human energetics properly, one needs also to understand all the subjective influences incorporated in the sense of 'effort'. Hence the reasons for this symposium and for its title.

Most of the research, so far, on human community energetics has been done in the USA, but there are in all other developed countries people working in human physiology, nutrition, anthropology, economics and demography, who if they choose to pool their skills and interests could make similar contributions. Neither money nor technology has been the principal limitation; rather, this has been the interest and the will. Multi-disciplinary approaches are, however, essential, and one purpose of the SSHB meeting was to bring possible collaborators together.

The first morning session, represented by the first four papers in this volume, was devoted to the physiology and psychology of the expenditure of energy and effort. Two are concerned with the difficult practical and conceptual problems of measurements (Nelms and Brown), one with a general review and evaluation of the determinants of energy expenditure (Cotes, Reed and Mortimer) and the fourth with disease factors, especially as found in a particular field study (Collins).

There follow five papers with a mainly energy-intake or nutritional orientation. Two of these (Rivers & Payne, and Weymes) are concerned with the complex problem of nutritional needs and the difficulty and even relevance of standards. The third is on the critical question of the conduct of dietary surveys, and we were particularly fortunate to have Dr. Anna Ferro-Luzzi from Rome, who has had so much experience of field surveys, as our contributor on this topic. Input-output studies are ambiguous in interpretation if no account is taken of stores and reserves, and this topic is reviewed in Chapter 8 (Norgan). Finally, in the nutritional section there was a contribution on some economic aspects (Bliss & Stern) which has far too rarely been considered by biologists.

A whole symposium could well have been devoted to social aspects, but the four contributors here well sample the sorts of issues that occur in cultural ecology. The first is devoted to the nature of work as seen through social anthropology (Wallman) and introduces a quite new and important perspective. This is followed by accounts of the ecological and economic strategies by both hunter-gatherers (particularly Australian aborigines (Harris)), and peasant cultivators (Richards) which well demonstrate the awareness of energy considerations in the economics of traditional peoples. But the inherent dangers of attempting to explain too much in cultural ecology in terms of such features as energy is cogently argued in the fourth contribution (Burnham).

The final three papers don't belong to any established camp: they focus totally on the ecology of energy. We were particularly fortunate to have as our second overseas speaker Dr. R. Brooke Thomas from the USA, who has pioneered human energy-flow studies. His contribution here (with McRae & Baker) is concerned with modelling these systems and their economic and demographic concomitants in the Andean altiplano. The last two papers (Bayliss-Smith and Spedding) deal with the analysis of energy systems at both the micro and macro levels, and especially in relation to the efficiency of agricultural systems. They show the way ahead.

The papers when presented were all followed by lively discussions, which added much to the value and enjoyment of the symposium. For this, the organizer thanks all the conference participants. He feels, however, that the spirit of this discussion could not be captured on paper even if the content could, and it was therefore decided not to try to publish it. Nevertheless, it is to be hoped that this publication will generate similar exchanges of view, for there can be little doubt that if it does, studies of energy and effort will prosper.

This book is dedicated to the memory of

JOSEPH SYDNEY WEINER

(1915–1982)

Founder of the Society for the Study of Human Biology

## MEASUREMENT OF WORK AND EFFORT
## PHYSIOLOGICAL ASPECTS

The Late J.D. Nelms

Army Personnel Research Establishment
Farnborough, Hants.

## 1. INTRODUCTION

It is well known that work physiologists are obsessed by the measurement of oxygen consumption. Why should this be so, when the muscle, in order to produce work, uses a variety of chemical substrate molecules in a complex enzymatic reaction, and oxygen is only the common oxidant?

Muscles contract by employing energy-rich phosphates to initiate contraction. Adenosine triphosphate (ATP) is the chief intermediary, and creatine phosphate is the chief immediate energy-rich store. But together these are only worth a few seconds of operation. Muscle glycogen can reconstitute these molecules without the need for oxygen but even its store is worth only a few seconds more. Thus, if no oxygen is available, metabolism is anaerobic and only about 10 seconds maximum work is available, as in an athletic sprint.

In normal exercise, oxygen is available and oxidation of substrates taken from the blood is possible. Therefore, a rational measurement of work should lie in quantifying the substrates – glucose and free fatty acids – which are used in a given time for a given task. Such simple molecules are easy to measure, but the site of measurement would have to be immediately before and immediately after each muscle (i.e., in the arterial supply and venous drainage of the muscle). Moreover, each individual muscle in a complex movement would have to contribute its share to the measurement. The technique would clearly be impossible to carry out.

Perhaps the levels of glucose in the arterial and venous blood could be compared and the quantity used in a given work activity deduced. The measurements are feasible but the results would be worthless, since the whole purpose of the homostatic system for glucose maintenance is to prevent blood level shifting from the normal. In the absence of a reliable measure of substrate usage, therefore, only the oxidant, oxygen,

is left as a common factor.    Fortunately, oxygen uptake is
easy to measure and its relationships to the chemical combus-
tion process are stoichiometric.    The relationship of oxygen
consumption to work is therefore linear, and fortunately so also
are the further relationships to heart function which are of immediate concern here.


2.    OXYGEN CONSUMPTION AND WORK

    During the past 40 years, great effort has been devoted to
the development of techniques by which to measure or to calculate the oxygen consumption associated with work.    The subject has been extensively and critically reviewed   (Åstrand &
Rodahl 1977, Banister & Brown 1968, Haisman 1967, Passmore
& Durnin 1955) and it may be helpful, by way of introduction,
to describe the latest in a series of attempts to standardize a
test procedure which could be employed simply and on a wide
scale as a screening test in various applications.
    The procedure in question has been derived in a NATO con-
text, for many armed forces have become increasingly concerned
over a suspected decline in the physical capacity of recruits
(reflecting general social changes), and also in men of a somewhat older group overtaken by the increasingly 'high technology-
low activity' nature of many military lifestyles.    The NATO
Defence Research Group fosters such collaborative projects between the member nations.
    Since the stationary cycle ergometer is now very widely avail-
able in laboratories, workplaces, and physical recreation and
health centres and (a matter of importance if results of reasonable accuracy are to be assured) can be readily calibrated, the
NATO test has employed this in preference to the traditional
laboratory treadmill.    The calibration method of Amor & Savill
(1973) may be used, and figure 1 shows the latest development
of this by Toft & Hopkinson (1981), recently completed at APRE.
    The subject begins pedalling at a workrate of 37.5 watts (a
setting of 0.5 kilopond on a 'Bodyguard' cycle pedalled at 75
revolutions per minute).    Each minute, a further load of 37.5
watts is added, by altering the brake band pressure, or its
equivalent on an electrically braked bicycle.    The heart rate
is monitored throughout by means of electrocardiogram (ECG)
leads on the chest, connected either directly or, with greater
convenience to the subject, by telemetry to an ECG recorder.
The endpoint for fit subjects is the maximum work level achieved,
as judged by inability to maintain pedalling rate, and this is

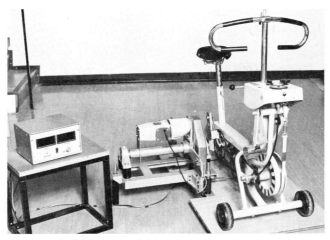

FIGURE 1.    Calibration of bicycle ergometer.


usually accompanied by a levelling off in the previously rising
heart rate.    Less fit subjects, or those in older age groups,
work to an endpoint heart rate of 190 minus their age in years.
In each case the final work rate is noted, as is the time taken
to reach the endpoint.

During the laboratory derivation and validation of the NATO
test, some of the levels assumed or calculated in the simplified
'practical' version given above were measured at the same time.
Thus, expired gas was collected in Douglas bags and was subsequently analysed for oxygen and carbon dioxide in paramag-
netic and infra-red gas analysers, after measurement of its
volume in a large (120 $\ell$) Tissot spirometer.    A sample of
blood was taken from an antecubital vein five minutes after
cessation of the exercise, to be analysed chemically for lactic
acid level.    The rate of oxygen consumption at maximal work
is expressed as the maximal oxygen uptake ($\dot{V}O_2$max) of the
subject.    Such studies typically result in values like those of
table 1, the lactic acid and heart rate levels giving confidence
that the effort has been close to maximal.

Tests of this kind used to assess maximum aerobic power
- that is to say, the capacity for prolonged work - are also
an immediate measure of what is called 'physical fitness'. More
important, they are reproducible measures for one individual
person, and the results alter predictably in response to physical fitness training - increasing by 5—15% (Rowell et al. 1964),

TABLE 1.    Typical values in maximal work.

| | Max. Heart Rate (BPM) | Lactic Acid (mmol/$\ell$) | Maximal Oxygen Uptake ($\ell$/min) | ((ml/kg)/min) |
|---|---|---|---|---|
| Men | 190 | 12 | 3.2 | 48 |
| Women | 193 | 12 | 2.2 | 37 |

and decreasing after bed-rest, semi-starvation, infection, de-hydration, and of course with age. $\dot{V}O_2$max, the maximal oxygen uptake, is thus a clear basic indicator of work capacity for the individual and, to a certain degree, also provides a means of comparing individuals. In addition, it permits the definition of work levels which are sustainable for increasingly long periods of time, up to and including a full working day or week. Moreover, since $\dot{V}O_2$max is a reproducible measure at any given time, a basis for the quantitative comparison of different jobs exists. This contention will now be examined.

## 2.1.   The comparison of individuals

It is obvious that, just as a horse can perform more actual work than a dog, and will use more oxygen in so doing, the same applies as between a big man and a little one, even when commonsense observation shows the small man to be in a higher state of 'physical fitness' than the large one. Body weight is thus a strong correlate of maximal oxygen uptake.

Body weight includes fat as a highly variable component. Bones and other body tissues form a further component, though one which is far less variable between individuals of the same general stature. Fortunately, it is easy to gain a relatively accurate estimate of body fat by measurement of skinfold thickness at standardized body sites (Durnin & Womersley 1974). Accurate laboratory methods are also available which make use of density measurements by underwater weighing, and of body water or lean body mass by isotopic methods (deuterium and potassium-40 respectively).

Subtracting the value for body fat thus obtained from the total body weight gives a figure for lean body mass which, although it includes variations in the weight component due to bones, liver, brain and other organs as between large and small individuals, can be a factor for standardizing the oxygen uptake of people of different type (and sex).

In summary, the most widely used measurement of physical working capacity may be expressed in the following ways:

$$\ell \; \text{min}^{-1} \; ... \; \text{for the given subject}$$
$$\text{m}\ell \, \text{kg}^{-1} \text{min}^{-1} \; ... \; \text{related to his body weight}$$
$$\text{m}\ell \, \text{kg}^{-1} \text{min}^{-1} \; ... \; \text{related to his lean body mass}$$

Two examples drawn from the literature (Myles & Allen 1980) demonstrate the pitfalls of ignoring body composition in $\dot{V}O_2$ max determinations - table 2.

TABLE 2.    Body weight and maximal oxygen uptake.

|  | Subject | |
|---|---|---|
|  | A | B |
| $\dot{V}O_2$ max ($\ell \, \text{min}^{-1}$) | 3.86 | 3.8 |
| 1.5 mile time (min) | 11.3 | 8.5 |
| 3.0 mile time (min) | 32.0 | 19.0 |
| Body weight (kg) | 98.0 | 64.5 |
| $\dot{V}O_2$ max ($\text{m}\ell \, \text{kg}^{-1} \text{min}^{-1}$) | 39.4 | 59.3 |

2.2.    Physiological basis for the correlation of heart rate, $\dot{V}O_2$ and work

The final common pathway acting as the limitation or chief determinant of maximal work performance is the system which delivers oxygen to the muscles.    In health, the limiting factor in this system is the performance of the heart.    There are circumstances in which the oxygen transfer into pulmonary blood from the air, or from the capillaries into the muscle cells, can be interfered with, but in health and at sea-level these oxygen-transfer mechanisms are adequate up to the maximum performance of the heart.   Thus, it is the cardiac output, expressed in litres of blood per minute, which governs the maximal rate of sustained work by an eventual limit on the oxygen supply to the muscles.

During exercise, a previously resting cardiac output of, say, 5 $\ell$ min$^{-1}$ may rise to about 30 $\ell$ min$^{-1}$.   At this time the blood vessels of the working muscles will be fully vasodilated and the proportion of the cardiac output which then flows through the muscles – by reason of this vasodilation coupled with major ad-justments in the remainder of the circulation – may typically rise to 85% from a resting value of about 15%.

Both components of the cardiac output – the heart rate, and the volume expelled in each beat (the stroke volume) – rise dra-matically in exercise under the influence of nervous impulses from the sympathetic nervous system, and blood-borne catecholamines from the adrenal medulla.   This greatly increased (indeed, neces-sarily equal) volume of blood returning to the heart is 'pushed' by the contracting muscles and 'pulled' by the increased thoracic and diaphragmatic movements of heavy breathing.

For example, heart rate may rise from about 70 bpm at rest and level off with a plateau at about 190 bpm when exercising maximally. Stroke volume approximately doubles, say from 70—140 m$\ell$ in a small sedentary man, or 85—190 m$\ell$ in a big athlete.   But stroke volume differs importantly from heart rate by coming to its plateau at lower levels of exercise, in the region of 40% of maximal.   Thus, heart rate continues to increase after stroke volume has reached its peak.

A recent study of current non-invasive techniques for cardiac output measurement in soldiers engaged in moderately active duties such as marching and digging, or severe physical exercise on an assault course, showed that it is not yet practicable to adapt so-phisticated clinical procedures for use in the field (Hanson 1981). Thus, cardiac output is, in these conditions, not a directly mea-surable quantity;  so neither is stroke volume.   It is therefore a fortunate circumstance that stroke volume should come to a pla-teau when exercise is only about 40% of what is possible.   For it follows that, above about 40% of maximal exercise, cardiac output is largely determined by heart rate alone, which of course is easily measured.

2.3.    The relationship of heart rate to oxygen consumption and work

It has been implicit in the foregoing arguments that heart rate,

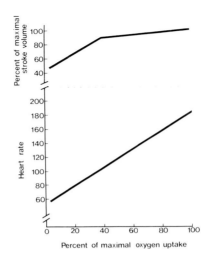

FIGURE 2.     Components of cardiac output, related to $\dot{V}O_2$max.
Redrawn from Åstrand et al. 1964.

the measurable parameter, has a well defined relationship to
work and oxygen consumption.    This relationship will now be
exposed.    Bearing in mind that the ability to produce and sustain muscular power is an all-important characteristic for a hunt-
ing-fleeing animal, it is a measure of the delicate complexity of
the interactive homeostatic mechanisms governing the relationship
that it turns out to be linear.    Power is produced by 'burning'
substrates in oxygen, which means that work and oxygen consumption are directly related.    Oxygen supply at the higher
rates is governed largely by heart rate (as shown above), which
means that oxygen consumption and heart rate are also directly
related.

These basic ideas are simple and the complexity of the control
mechanism required would be expected to produce a great deal of
'noise' in the system, and hence a wide variability of values around
the fundamental relationship.    In fact, whilst there is wide variation between individuals which has to be handled statistically,
as is usual in biological samples, the results from a single individual are reproducible.    Figure 3 (from Lange Andersen et al.
1978) makes these relationships plain.

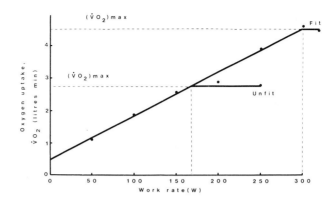

FIGURE 3.    Relation between oxygen uptake ($\dot{V}O_2$) and work
rate (W) in fit and unfit subjects.


## 2.4.    Prediction from submaximal tests

For an individual subject, the linear relations between heart
rate, oxygen consumption and workload implies that,  once the
slope of the line has been established by measurement (say, on
a calibrated bicycle ergometer), any value of oxygen uptake and
work can be deduced.    There is then no need to expose the
subject to the stress of developing a maximal heart rate while
being tested, and this is helpful when the subject is unfit  or
vulnerable to the undesirable effects of unaccustomed maximal
exercise.    In addition, it has already been hinted that the maximal oxygen uptake turns out to be a remarkably constant value
for the individual in a given state of fitness and age, and also
has validity in the assessment of the energy cost of particular
work tasks.    Thus, $\dot{V}O_2$max is a desirable figure to establish
and, because its accurate measurement is time-consuming and
needs laboratory equipment, a series of predictive methods has
been developed.

The predictive methods are based on population studies relating actual heart rate and oxygen uptake at levels up to the
true $\dot{V}O_2$max and the true maximal heart rate of the individuals
making up the population.    The meaning of maximal heart rate
must be strictly defined and adhered to - in essence it corresponds to the plateau beyond which heart rate does not increase,

though higher work rates may be possible for a very short
time by using up the remaining anaerobic reserves.
Knowing this, the submaximal heart rate and $\dot{V}O_2$ of an individual can be measured and extrapolated to the population
maximal heart rate, and the maximal oxygen uptake read across.
The oxygen consumption measurement may be omitted altogether
by measuring work rate and heart rate (in a step or bicycle
test) at several submaximal levels.     Measurement of work and
heart rate at only one level is a final simplification.

In all these procedures, two major corrections must be made
– for age and sex (Hermansen 1973).    The widely published
equations and nomograms (Åstrand & Ryhming 1954) include
these factors.

As the predictive procedures were progressively simplified
in the manner described, the data lost the validity and repro-
ducibility of the direct treadmill measurement, and a progressively greater variance accrued to them.     Rowell et al. (1964)
and Davies (1968) have emphasized these limitations.     But, by
choosing the test method appropriate to the aim of the study and
by using great care in the interpretation of individual results,
the submaximal tests have proved invaluable in work measurement.


## 2.5.    Physiological endpoints of work

When work must be done, it is the chosen rate of energy expenditure which determines the total quantity of work which can
be performed.    The endpoint, if there is one, will be imposed
by exhaustion or fatigue – the latter a generalized 'feeling' of
gradual onset, or a more acute malaise of specific muscle groups
or overworked musculo-skeletal structures.    Exhaustion and
fatigue are very difficult terms to define, since the subjective
awareness of them – which is all that counts to the sufferer –
is related by no means clearly to the local causes of those sensations.    We have no words for rising lactic acid levels or falling pH in blood or muscle, nor even a clear perception of inadequate venous return, or of a heart rate which can rise no further.    Lacking analogues of the rudimentary instruments for
monitoring the electrical and mechanical reserve health of our
car engine, we must rely on strong but imprecise sensations of
'feeling too hot' or 'gasping for breath' as signals of impending
exhaustion, and vague alterations of proprioceptive sensation to
warn of fatigue.

In the laboratory, objective measures are much easier to make,
though not necessarily more informative about the underlying

physiological state.    Thus, the moment when a subject can no longer match his pedalling rate to a metronome may be easy to time.    Even so, the physiological determinants of the endpoint are far from fully worked out, and the calm objective appraisal is in marked contrast to the picture of determination and distress seen in the face of a well motivated subject at such a moment.

It is reasonable to suppose that, if the oxygen supply system is the general limitation to this kind of work, the place at which a deficiency of oxygen will act is the working muscle itself, and the 'off' signal will be generated in the muscle, even though the final cut-off switch may be cerebral.    Among the possible factors are:    conduction at the neuromuscular junction;    muscle-fibre conduction and contraction mechanisms;    energy-rich phosphate store depletion;    rising lactic acid levels affecting pH-sensitive enzymes;    and diminished substrate uptake from blood (glucose and free fatty acids).

Interacting with several of these is muscle glycogen depletion, which has been shown to be almost total in exercise to exhaustion (Maehlum & Hermansen 1978).    Studies showing the direct relationship of muscle glycogen depletion with lactic acid level and its course of recovery in different experimental regimes (Hermansen & Vaage 1977) have had interesting implications for the manipulation of glycogen stores in endurance athletes, and for military groups (Hodgdon et al. 1978).

2.6.    Effects of training on work

In the realistic assessment of work it is important to take account of the influence exerted by training – a word which implies skill as well as cardiovascular or muscular development. For example, even the 'simple' act of bicycling may be quite unknown to isolated village subjects.    And the fact that the mean capacity of US women to perform chin-ups is less than 1.0 has serious implications for the use of this test in survey work!

Once the all-important skill components have been separated out, and the tests have been selected to be possible for all subjects, the effect of specific training may always be demonstrated. Table 3, from the work of Kowal et al. (1978) shows an example from male and female US Army recruits before and after basic military training.

TABLE 3.    Fitness and strength before and after training.

| | | Work Response | | Isometric Strength | |
|---|---|---|---|---|---|
| | | HR max (BPM) | $\dot{V}O_2$ max (m$\ell$ kg$^{-1}$min$^{-1}$) | Trunk Extension (kg) | Leg Extension (kg) |
| Men | Pre | 190.6 | 50.7 | 72.3 | 142.5 |
| | Post | 184.7 | 51.6 | 78.0 | 156.5 |
| Women | Pre | 190.7 | 36.9 | 47.6 | 91.4 |
| | Post | 184.6 | 38.8 | 56.1 | 105.6 |

## 3.    THE CAPACITY FOR SUSTAINED WORK

Fully anaerobic work, in which virtually none of the consum-
able components can be replenished from outside the muscle, can
last but a few seconds.    Progressively less intense work-rates
permit proportionately greater endurance – for minutes, for hours
or finally for the typical intermittent social pattern of daily physical labour without exhaustion.    Such normal long-term patterns
(and their now even more customary sedentary variants) are
quite unsuitable for transfer into the laboratory for precise measurement.    Nor are the laboratory methods so far discussed suitable for their measurement at the workplace.    Some sacrifice of
precision must be accepted in the search for approaches and techniques which can be made compatible with daily work – not only
from the viewpoint of social acceptability, but also to avoid interference with the very work whose assessment is sought.    Some
of these techniques will now be mentioned.

### 3.1.    Oxygen consumption at the workplace

Several methods are now available for measuring $\dot{V}O_2$ during
normal work by means of gas analysis or gas collection in an apparatus small enough to be carried conveniently.    The Oxylog
(Humphrey & Wolff 1977) measures volume, and inspired and

expired $O_2$ levels, in air breathed through a mask. The 'Miser' (Eley et al. 1976) measures flow and takes a small sample of expired air for laboratory analysis later.

Among these techniques, the KM respirometer (Kofranyi & Michaelis 1940) has been the most widely used and validated, and is to be found pictured in all reviews on the subject. The moving-diaphragm gas meter box may be worn on the back and, as well as measuring the expired gas flow, the KM diverts a small sample into a bladder which is removed for separate analysis using standard laboratory methods. In common with the Oxylog, the KM is limited in its capacity to handle very high gas flow, and respiration becomes increasingly obstructed when more than about 50 $\ell\,min^{-1}$ are expired. Furthermore, the size of the collecting bag limits the duration of a measurement to about 25 minutes at high rates of flow. Nevertheless, because of its robustness and suitability for field work in rigorous 'expedition' situations, the KM respirometer has encouraged development. Haisman (1970) increased the collecting bag capacity to provide 60 minutes duration, and improved the acceptability of the apparatus by substituting a mask for the uncomfortable mouth-piece and nose clip. Vogel et al. (1973) were able to make measurements on the British Biathlon team during 17 km ski 'races' involving ventilation rates up to 120 $\ell\,min^{-1}$ by connecting two of the meters in parallel on a small back-pack.

Amor et al. (1976) have provided a detailed account of the means by which the KM may be modified to work in arctic field experiments down to -30° C. Figure 4 shows one example of a task assessment – movement over snow – by these techniques, and shows how practical conclusions may be drawn from the data. In this case it may be seen that skiing, a task taught with difficulty and expense, is no more efficient than easily learned snow-shoeing up to a speed of 1 m $s^{-1}$ (2.2 mph).

## 3.2.   Standardization of work capacity

When, by the foregoing techniques, the energy cost of a given task is known for a particular individual, and his or her maximal oxygen uptake has been measured or predicted, the work can be described as requiring a precise percentage of his maximal oxygen uptake ($\%\dot{V}O_2max$). While this expression cannot be used to standardize a new work task or rate of work without knowing the characteristics of the population available to perform it, there is obviously value in the possibility of predicting the proportion of people who should be able to accomplish it with ease.

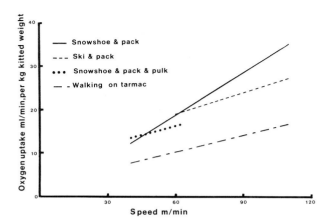

FIGURE 4.    Energy cost of skiing, snowshoeing and walking.

Many surveys have shown that the acceptable duration of work is related to individual cost expressed as $\%\dot{V}O_2$max.    The studies of Åstrand (1960), and Åstrand <u>et al</u>. (1959) examined work at 50% $\dot{V}O_2$max, for one hour, in four men and in 18 females over 40, culminating in an heroic study of four subjects aged between 20 and 40, lasting eight hours.    Bicycle and treadmill were alternated, and the subjects rested for ten minutes each hour.    From the fact that physiological changes and ill-effects were minimal, it has been concluded that the 50% level is the highest which should be attempted for an eight hour working day.

However, it is uniformly accepted that this would be far too high a figure for social planning purposes except in the case of elite trained groups such as lumberjacks.    Astrand & Rodahl (1977) themselves suggested 30-40% $\dot{V}O_2$max as the limit if fatigue is not to be experienced after an eight hour day.    Muller (1962) recommends a daily level of about 20%, based on measurements of rising pulse rates.

An example may serve to underline the point of the $\%\dot{V}O_2$max concept.    Suppose two men of about the same body weight need to shovel snow from their driveways.    This is about 1.5 $\ell$ min$^{-1}$ oxygen uptake task (in the 'vigorous' category in our present social climate).    One man has a $\dot{V}O_2$max of 4.5 $\ell$ min$^{-1}$.    He will comfortably extract his car, even if he must shovel all day, for he is working at 33% $\dot{V}O_2$max. The other, having a low oxygen uptake of 2.0 $\ell$ $min^{-1}$, no doubt the product of easy living, would have but 220 minutes (Knuttgen 1977) before his blood lactate and muscle glycogen reached their limits. He would indeed be unwise to go so far, bearing in mind his low level of physical fitness. At any rate, his 75% $\dot{V}O_2$max work rate inexorably limits the endurance available to him. Similar examples could be devised to show even more dramatic results, for at 100% $\dot{V}O_2$max only 2—10 min will pass before exhaustion, or 20—30 min at 90% $\dot{V}O_2$max in athletic subjects.

3.3. Muscle strength measurement and % maximal isometric power

Techniques for the measurement of muscle groups working in their normal relationships are less well developed and standardized between laboratories than are the aerobic measurements detailed above, although increasing attention is now being given to them by the NATO research study group mentioned above. Hermansen (1974) shows the dynamometers, strain gauges and amplifiers used for the measurement of arm flexors, abdominal and back muscles, and knee extensors in isometric contraction.

It is of interest that isometric contraction of muscles also has a time course which can be strictly related to their % maximum contraction. The endurance of such contractions will be measured in seconds at about 75% of maximum. About one minute is possible at 50%, and four minutes duration is available when the muscle is contracting steadily at 25% of its maximum tension. The time asymptote of this curve occurs at about 15% max, though this should not be interpreted as an infinite capacity for sustained contraction.

Clearly, the difference between % maximum isometric contraction of a specific muscle and % maximum oxygen uptake of a person performing aerobic work is related to the fact that, in normal movements, isometric contractions play a small part and muscles are allowed to relax between active thrusts. Bicycling or walking are familiar examples and, in such cases, a tension x time factor is the operative one.

4. OTHER MEASURES OF WORK

Heart rate has been used as an index for the 'heaviness' of work (indirectly, the energy cost), and also for the stressful aspects by using it as a measure of physiological strain. Clearly, the matter is complicated by the fact that heart rate is notoriously under the influence of emotional, endocrine and pharmacological factors, and in part therefore belongs to the domain of 'effort'.

Heart rate has been examined beat by beat (with an almost insoluble problem of data analysis), or integrated over minutes, hours, or during a longer work period.    Rodahl (1977) has re-viewed the value of this approach using small portable tape re-corders triggered by the QRS complex of the electrocardiogram, and modified to run for up to eight hours.    Fairly good agree-ment with $VO_2$ was reported.    Agreement was also found with the attempt to derive a heart rate analogue to the % maximum concept, with a correction for the highly variable resting heart rate of different individuals.    In absolute heart rate terms, how-ever, Rodahl opted for a classification of prolonged work as fol-lows:

| | | |
|---|---|---|
| Light work | Up to 90 | BPM |
| Moderate work | 90—110 | BPM |
| Heavy work | 110—130 | BPM |
| Very heavy work | 130—150 | BPM |
| Extremely heavy work | 150—170 | BPM |

Other methods of estimating the total number of heart beats during a shift are reviewed by Lange Andersen et al. (1978). Those which use the ECG pulse to trigger electrochemical plat-ing of a 'battery' whose discharge can subsequently be quanti-fied, and those triggering the escapement of a digital watch are of interest, but all suffer from grave inaccuracies due to elec-trode and EMG artefacts.    The final values may thus contain variance sufficient to swamp the differences between work levels, at least for comparisons of any rigour.

The Medilog recorder (Oxford Instruments Ltd, McKinnon 1974) is a small battery-operated multi-channel tape cassette recorder designed for continuous operation up to 24 hours.    It can be at-tached to a belt, allowing an individual to go about his normal duties unencumbered by the leads usually associated with physio-logical monitoring.    Most commonly to date, electrocardiographic signals have been recorded for subsequent analysis of heart rate following appropriate electrical amplification and integration using

a fast replay unit. It is also possible to record other physiologically derived signals such as blood pressure, skin and body
core temperatures, oxygen uptake and ventilation from the Oxylog, $O_2$ and $CO_2$ concentrations and ventilation from the Miser,
and accelerometer or pedometer outputs as indicators of physical
activity.

### 4.1. Energy intake and activity measurements

So far, attention has been paid only to the oxygen side of
the energy equation, the resultant work being expressed as applied load and duration, or as power output, or as their oxygen
equivalent. But endurance work must ultimately be related to
energy intake and, when looking from this direction, it is convenient to employ work units appropriate to the calorific content
of food.

A great body of work exists in this area, and values for almost every kind of occupation are given by Passmore & Durnin
(1955). Apart from the standard techniques for oxygen consumption mentioned above, the methods of study have included
diary techniques for activity records, and food intake studies
making use of standard calorific values for each item of food consumed. Such studies may be carried out under very close control in metabolic isolation suites, but the highest accuracy to be
had during survey work in a social context is about ± 10%.

Some methods seek to quantify jobs in terms of the movements
made rather than the relation to energy cost per se. The inertia
pedometer is the simplest, moving a counter when a given acceleration is sensed, but over-estimating when complex movements
in several planes are performed, or resonances occur. It is
only suitable for stereotyped tasks, such as plain walking. Self-
winding watch mechanisms have been used to register acceleration
and duration of movements, also by inertia.

Among the ergonomically orientated measurements – which seek
to reduce the workload by improvements in design – there are mechanical, photographic, opto-electronic, and computer methods
(Holzhausen 1980). Mechanical methods are best suited to the
assessment of reach-envelopes, and provide a reading for each
position attained in space by different parts of the body during
work movements. Photographic methods have received much development attention in slow-motion and time-lapse refinements, as
well as the provision of standardized backgrounds and stereo-
photography for still work. All the advanced techniques of the
modern TV producer could be employed with good effect in this

field, and one wonders whether laser holography could not provide the ultimate in stereophotogrammetry, permitting the observer to 'move around' the hologram in the act of measurement. Finally, in this group, the classical time-and-motion approach may be mentioned, in which experienced observers follow workers in order to note, in minutes and seconds, the exact operations performed. To be accurate and well correlated with the physiological results (and it can be) the method is highly demanding and labour-intensive.

## 4.2.   Electrophysiological techniques

Table 4, from Rau (1980) summarized the remaining, largely non-invasive, methods of this type which have proved useful in the assessment of various aspects of work. However, with the exception of the ECG complex used as a trigger for heart rate totalizers, described above, none has been quantified in such a way as to measure the capacity for sustained work.

## 4.3.   Lifting work

A significant component of all human manual work consists in lifting and handling loads, in mechanically developed societies no less than in the agrarian world. The musculo-skeletal system tends customarily to be fairly close to its feedback limits, perhaps because of its adaptability and regenerability. In other words, Homo sapiens (sedentarius) will atrophy to about the muscular strength appropriate to his work needs just as well as H.sapiens (athleticus) will hypertrophy. But both tend to operate rather near the 'catastrophe cusp' of musculo-skeletal damage, and proprioceptive signals about the reserves remaining in hand seem difficult for man to interpret. Measurement of work in a social context should therefore include attention to lifting activities.

A recent review by Ayoub (1980) has delineated three approaches. By encouraging deliberate introspection, the loads and repetitive lifts acceptable to wide ranges of subjects have been analysed. In other experiments heart rate, energy consumption, or $\dot{V}O_2$ have been measured; the last so as to derive a $\%\dot{V}O_2$max relationship. A third, more speculative, approach attempts the integrative modelling of proportional forces acting on the joints involved in specific lifts. At present it would appear that there is a significant sex difference in lifting capacity, but not an age difference. The physical characteristics of the lift (size, weight),

TABLE 4.   Electrophysiological methods.

| Parameter and measuring technique | Principle measurement range (max. amplitude) | Signal frequency range (Hz) | Electrode type |
|---|---|---|---|
| Electrocardiography (ECG) | 5 mV | 0.1–250 | skin |
| Electroencephalography (EEG) | 500 µV | 0.1–40 | skin scalp |
| Electromyography (EMG) | 10 mV | 1–2000 1–10 000 | skin needle wire |
| Nerve potentials (ENG) | 5 mV 5 mV | 1–2000 1–10 000 | skin needle wire |
| Eye potentials EOG ERG | 5 mV 1 mV | dc–100 dc–100 | skin contact |
| Galvanic skin response (GSR) | 500 kΩ | 0.01–10 | skin |
| Visual evoked | 100 µV | 0.1 –40 | skin scalp |

and height and frequency of lift, have linear effects on capability.

Approaching the problem from the direction of potential back strain, the pioneering work of the Surrey University group established detailed lifting criteria based on measured intra-abdominal pressure levels (Davis & Stubbs 1977), and work at APRE is extending this to questions of repetitive fatigue and lifting in awkward spaces.

## 4.4. The questionnaire method of work assessment

Questionnaire techniques have been widely used to supplement direct measurements at the workplace. They fall into two categories, according to whether the information is to be recorded in 'real time', or retrospectively.

Concurrent or diary questionnaires aim to account for all the period covered, and no minute of the day should escape into a 'not-categorized' section. The activities listed must necessarily be simplified if co-operation is to be assured, especially when the duration is a week, the period for which the most reliable results are likely to be forthcoming. In some studies, food intake over the period has also been assessed as a cross-check. The diary technique can be used for short periods of work with a minute-by-minute categorization, but at some point the interference with the task itself makes an observer essential - when the technique really becomes time-and-motion study.

The recall questionnaire looks back to an earlier period of work, though ideally this should be only a few days away. Its chief use has been in the domain of social epidemiology. The Framingham study of 24 hours in the life of 5127 subjects (Kannel 1967, Kannel & Gordon 1974) is perhaps the classic; however, I cannot but give more attention to that of Morris' group, for it concerns Civil Servants. Some 17 000 office workers recalled two day's activities of the previous week during 1968—1970, accounting for each five minute period. By 1980, over 1100 had suffered their first coronary heart disease episode, and it could be shown that the group classified as vigorous exercisers had a 3.1% incidence, compared with the 6.9% rate of the rest (Morris et al. 1980).

Thus, such questionnaires provide definite information of social value, and yet this information is of necessity derived from categories of work too broad to possess any obvious precision (table 5). The solution to the paradox lies in the enormous numbers employed. If recall questionnaires are considered for more detailed work tasks or smaller numbers, it must be borne in mind that subjects have a tendency to over-estimate the amount and intensity of activity (Lange Andersen et al. 1978).


TABLE 5.    Activity levels by questionnaire.

| | |
|---|---|
| Sedentary | Reading.   Desk work. |
| Light | Car driving.   Washing up. |
| Moderate | Walking.   Shopping. |
| Active | Lawn mowing. |
| Vigorous exercise | Sports.   Snow clearing. Climbing 500 stairs/day. |


## 5.   THE MANY CONTEXTS OF WORK

Each disciplinary approach tends to study those factors in its own domain which most strongly affect work performance. The ergonomist concentrates on the relationship of modifiable aspects of the task and the characteristics of the available human population.   The social scientist identifies the major and minor social contexts of the workplace, and the attitudes to it, as among the primary factors.   The exercise physiologist focuses most upon the body – its size, muscularity, training, age and sex, and the condition of its cardiovascular and musculoskeletal systems.   All these disciplines can show that 'their' context may have overwhelming effects on work performance. The environmental physiologist too can display his share of major determinants which condition work;   among them heat, cold hypoxia, vibration, noise, and sleep deprivation.

The responses of heart rate and body temperature to work in the heat are well understood, and the encumbrance effects of work in a cold environment are easy to measure.   Much less well quantified are the responses to vibration and noise per se, but sleep deprivation has been shown, perhaps surprisingly, to have no effect on physical as opposed to mental performance (Haslam 1981). The graded increase of noradrenalin with exercise of increasing intensity is well known, but particularly clear responses in blood hormone levels are now also being described, with the availability of adequate and convenient analytical techniques.   Aakvaag et al. (1978) investigated officer cadets during five days of intense activity, including 30 km night marches, using 46 MJ/day of energy

against an intake of 6.3 MJ/day and getting only 2.5 hours
sleep overall.    Figure 5 shows powerful responses to growth
hormone and testosterone level to this stressful regime of hard
work.

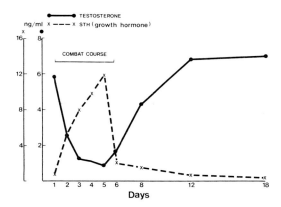

FIGURE 5.    Hormone changes in response to five days intense
activity.

## 6.    THE CONCEPT OF EFFORT

'Work' has proved to be a portmanteau term requiring clear
definition to avoid diffuse and general treatment.    'Effort', the
other theme of this book, will offer no less a temptation towards
generality, for it carries very clear connotations of attitude and
internal mental perceptions, as opposed to physiological values.
In physiological terms the only safe definition may lie in the
engineering domain – 'effort' is to be measured as physiological
strain.    Thus, the general cardio-respiratory response to work
in increased heart rate and $\dot{V}O_2$ is a measure of effort. But the
perceived effort is a more complex function of these changes, for
it will vary according to fitness, age and sex.    Specific strain
may also be seen when muscles, joints and spinal structures approach their limits.
Moreover, there are changes which are not perceived directly
as such, for example, the testosterone levels mentioned above.
It is hard to know how to place these in terms of 'effort', yet
clearly they are a strain, a response to stress.    By contrast,
fatigue is conscious, and yet the objective concomitant factors

are not very clear.    Lactic acid level in muscle may be causally associated with the feeling of acute fatigue, but lactic acid level in blood is not the sole reason, if it is a reason at all, for end-of-shift tiredness.

In short, what is perceived may not be measurable and what can be measured may not be perceived.    Thus, physiology has as yet little to offer in the realm of effort, and it is significant that various laboratories well known for their objective studies of cardiac output, $\dot{V}O_2$ max and the like, have been paying attention to the Relative Perceived Exertion concepts of Borg (1971). The RPE scales and their analogues have proved to be quite robust, and have aided new thinking by their ability to include feelings, for which there are only associative descriptors of a 'poetic' vagueness, as well as stress responses for which the corresponding strains are as yet unknown.

Above and beyond this, as the world moves onward from a state of generalized labour in order to survive, the old balance between a sense of effort and a fear of starvation now becomes replaced by social preconceptions as perhaps the strongest determinants of work and effort.


REFERENCES

Aakvaag, A., Fonnum, F. & Opstad, P.K., 1978, Hormonal changes in serum in young officer cadets during prolonged military activities.    NATO symposium on physical fitness with special reference to military forces.    NATO DS/DR (78)98,p.107.

Amor, A.F. & Savill, M.W., 1973, The calibration of bicycle ergometers.    Technical Memorandum 45/73, Army Personnel Research Establishment, Farnborough, Hants.

Amor, A.F., Worsley, D.E., Jenkins, F.T., Gooderson, C.Y. & Kerry, C.W.M., 1976, Energy expenditure in military skiing. Report 47/75, Army Personnel Research Establishment, Farnborough, Hants.

Åstrand, I., 1960, Aerobic work capacity in men and women with special reference to age.    Acta Physiologica Scandinavica 49, Supp. 169, 1-92.

Åstrand, P.-O. & Rodahl, K., 1977, Textbook of Work Physiology (New York: McGraw Hill).

Åstrand, P.-O. & Ryhming, I., 1954, A nomogram for calculation of aerobic capacity (physical fitness) from pulse rate during submaximal work.   Journal of Applied Physiology 7, 218-
221.

Åstrand, I., Åstrand, P.-O. & Rodahl, K., 1959, Maximal heart
rate during work in older men.   Journal of Applied Physiology
14(4), 562-566.

Åstrand, P.-O., Cuddy, T.E., Saltin, B. & Stenberg, J. 1964,
Cardiac output during submaximal and maximal work.   Journal
of Applied Physiology 19(2), 268-274.

Ayoub, M.M., 1980, Physical work limits for lifting tasks.   In
Manned Systems Design: New Methods & Equipment, Vol. 1
(NATO Defence Research Section)   (To be published by Plenum
Press), pp. 219-250.

Banister, E.W. & Brown, S.R., 1968, The relative energy requirements of physical activity.   In Exercise Physiology, edited
by H.B. Falls   (New York: Academic Press); pp.267-322.

Borg, G., 1971, The perception of physical performance.   In
Frontiers of Fitness, edited by R.J. Shephard (Springfield,
Illinois: C.C. Thomas), pp.280-294.

Davies, C.T.M., 1968, Limitations to the prediction of maximal
oxygen intake from cardiac frequency measurements.   Journal
of Applied Physiology 24(5), 700-706.

Davis, P.R. & Stubbs, D.A., 1977, Safe levels of manual forces
for young males.   Applied Ergonomics 8(3), 141-150.

Durnin, J.V.G.A. & Womersley, J., 1974, Body fat assessed
from total body density and its estimation from skinfold thickness:
measurements on 481 men and women aged from 16 to 72 years.
British Journal of Nutrition 32, 77-97.

Eley, C., Goldsmith, R., Layman, D. & Wright, B.M., 1976, A
miniature indicating and sampling respirometer (Miser).   Journal
of Physiology, London, 256, 59-60.P.

Haisman, M.F., 1967, A review of physical fitness tests with particular reference to the maximal oxygen intake.   Research Memo-
andum P3, Army Personnel Research Establishment, Farnborough,
Hants.

Haisman, M.F., 1970, The energy expenditure of jungle patrols.
Research Report 1/70, Army Personnel Research Establishment,
Farnborough, Hants.

Hanson, G., 1981.    In Non-Invasive Measurement of Cardiac
Output - a Consultative Study, edited by W.I. Hopkinson
(APRE Memo 80M503).

Haslam, D., 1981, The military performance of soldiers in continuous operations.    In The 24-hour Workday:  a Symposium of
Variation in Work-Sleep Schedules    (National Institute for Occupational Safety & Health), in the press.

Hermansen, L., 1973, Oxygen transport during exercise in human subjects.    Acta Physiologica Scandinavica, Supplement 399,

Hermansen, L., 1974, Individual differences.    In Fitness,
Health and Work Capacity, edited by L. Larson    (London:
Collier MacMillan), pp.295-419.

Hermansen, L. & Vaage, O., 1977, Lactate disappearance and
glycogen synthesis in human muscle after maximal exercise.
American Journal of Physiology 2(5), E422-429.

Hodgdon, J.A., Goforth, H.W. & Hilderbrand, R.L., 1978,
Carbohydrate loading and endurance in naval special warfare
personnel.    In NATO Symposium on Physical Fitness with Special Reference to Military Forces, NATO DS/DR (78)98 pp.115-
123.

Kolzhausen, K.-P., 1980, Analyses of human movements for work-
place design.    In Manned Systems Design: New Methods and
Equipment, NATO Defence Research Section, Vol II, pp.556-587
(To be published by Plenum Press).

Humphrey, S.J.E. & Wolff, H.S., 1977, The Oxylog.    Journal
of Physiology, London, 267,12P.

Kannel, W.B., 1967, Habitual level of physical activity and risk
of coronary heart disease: The Framingham Study. Canadian
Medical Association Journal    96, 811-812.

Kannel, W.B. & Gordon, T., 1974, Obesity and cardiovascular
disease: The Framingham Study.    In Obesity, edited by W.
Burland, P. Samuel and J. Yudkin    (London: Churchill Livingstone), pp.24-51.

Knuttgen, H.G., 1977, Physiological factors in fatigue.    In Phy-
ical Work and Effort.    Wenner Gren Centre International
Symposia Vol. 28. Edited by G. Borg (Pergamon).

Kofranyi, E. & Michaelis, H.F., 1940, Ein tragbarer Apparat zur
Bestimmung des Gasstoffwechsels.    Arbeitsphysiologie, Interna-
tionale Zeitschrift für die Physiologie des Menschen bei Arbeit
und Sport  11, 148-150.

Kowal, D.M., Vogel, J.A., Patton, J.F., Daniels, W.L. & Sharp, D.S., 1978, Evaluation and requirements for fitness upon entry into the US Army. In NATO Symposium on Physical Fitness with Special Reference to Military Forces, DS/DR (78)98, pp.93-98.

Lange Anderson, K., Masironi, R., Rutenfranz, J. & Selinger, V., 1978, Habitual physical activity and health. WHO Regional Publications European Series Number 6 (Copenhagen: WHO).

Maehlum, S. & Hermansen, L., 1978, Muscle glycogen concentration during recovery after prolonged severe exercise in fasting subjects. Scandinavian Journal of Clinical Laboratory Investigation 38, 557-560.

McKinnon, J.B., 1974, Miniature 4-channel cassette recorder for physiological and other variables. Biotelemetry II. 2nd International Symposium, Davos, edited by P.A. Neukomm (Basel: S. Karger).

Morris, J.N., Everitt, M.G., Pollard, R., Chave, S.P.W. & Semmence, A.M., 1980, Vigorous exercise in leisure time: protection against coronary heart disease. The Lancet, 1207-1210.

Muller, E.A., 1962, Occupational work capacity. Ergonomics 5, 445-451.

Myles, W.S. & Allen, C.L., 1980, Personal communication.

Passmore, R. & Durnin, J.V.G.A., 1955, Human energy expenditure. Physiological Reviews 35, 801-840.

Rau, G., 1980, Electrophysiological measurement techniques. In Manned Systems Design: New Methods and Equipment, NATO Defence Research Section, Vol. I (To be published by Plenum Press), pp.181-198.

Rodahl, K., 1977, On the assessment of physical work stress. In Physical Work and Effort, Wenner Gren International Symposia, Vol. 28, edited by G. Borg (Pergamon).

Rowell, L.B., Taylor, H.L. & Wang, Y., 1964, Limitations to prediction of maximal oxygen intake. Journal of Applied Physiology 19(5) 919-927.

Toft, R.J. & Hopkinson, W.I., 1981, An electrical calibrator for bicycle ergometers. Memo 81A, Army Personnel Research Establishment, Farnborough, Hants.

Vogel, J.A., Amor, A.F., Crowdy, J.P., 1973, Physiological investigations on the British Biathlon Team. Report 10/73, Army Personnel Research Establishment, Farnborough, Hants, pp.1-34.

# MEASUREMENT OF MENTAL EFFORT:
## SOME THEORETICAL AND PRACTICAL ISSUES

I.D. Brown

MRC Applied Psychology Unit, 15 Chaucer Road,
Cambridge CB2 2EF

## 1. INTRODUCTION

As Chiles & Alluisi (1979) have recently pointed out, psychologists still have no generally accepted definition of the term 'workload'. It follows that there will also be little agreement on the definition of the term 'mental effort'. Thus, it is hardly surprising that psychology has lagged behind certain other disciplines contributing towards ergonomics, such as anatomy and physiology, in the provision of practical guidelines on parameters of work design.

This shortfall has become increasingly important in recent years, at least for the more developed countries, as more and more jobs demand mental rather than physical effort from the workforce. It is particularly unfortunate for the design of certain highly specialized jobs, such as air traffic control, chemical process control, or piloting civil aircraft, where the consequences of even apparently simple errors can occasionally be disastrous (e.g., Roitsch et al. 1977). However, even quite mundane jobs may produce industrial disturbances, or psychosomatic disorders among the workforce, if the mental effort required by their tasks is inappropriately designed, being set either too high or too low for optimal job satisfaction (Sell & Shipley 1979).

This problem of mental workload and effort measurement has been recognized for many years. Why then have psychologists been relatively unsuccessful in producing practical solutions? The difficulty here appears to be threefold:

(a) Mental load is multidimensional. The demands of a task may be sensory, perceptual, cognitive, attentional, or perceptual-motor. These demands will vary over time, overlap temporally, and sometimes compete for the same processing resources. Therefore, a simple analytic approach to the evaluation of task demands is often unreliable if the job is at all complex.

(b)    As Bainbridge (1974) has pointed out, the mental effort devoted to a task and the performance level achieved on that task are discontinuous functions of the task's demands.    She goes on to suggest that it is, in fact, difficult even to define the terms independently.    The problem here, of course, is that operators can very often use alternative strategies to produce equally acceptable performance for different levels of mental effort.  For example, they may recode incoming information, alter their visual search patterns, or modify their speed-accuracy trade-off function (Welford 1978).    An additional and equally obvious difficulty here is that individuals differ widely in their capacity for processing information.    Thus, the correlation between mental load and mental effort will differ among individuals, as well as between tasks and over time (Dornic 1977).

(c)    The extent to which task demands are met is a function of the worker's level of motivation (e.g., Welford 1974).    An inadequately motivated worker will usually fail to produce the mental effort required by task demands.    This may impair the quality, or the quantity of output, or both.    At the other extreme an overmotivated worker may expend too much mental effort in performing certain components of a task, to the detriment of others.    The complication here is that motivation is itself a function of mental load and effort, as well as being determined by other job characteristics such as pay and working conditions.

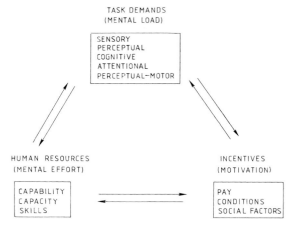

FIGURE 1.    Mental effort is a complex interactive function of multidimensional task demands, motivational factors associated with the job, and individual operator characteristics.

The interdependence between task demands, human resources
and motivation is illustrated in figure 1.   This figure may also
be used to suggest an historical explanation for psychology's
apparent failure, so far, to cope adequately with mental load
and effort measurement.   The relationship between task demands
and human processing resources has largely been addressed by
experimental psychologists, who have in fact made considerable
progress in both theory and methodology.   The relationship
between job demands and motivation, on the other hand, has
largely been addressed by occupational and social psychologists,
who also have made considerable progress in theory and methodology.   However, it is only in fairly recent years that the concepts developed within these different branches of psychology
have been integrated in a manner which permitted applied research on mental workload measurement to advance significantly.
This seems to have come about through the increased recognition
by society of the need for collaborative research on job satisfaction and quality of working life.

   These, then, are some of the psychologist's general difficulties
in this field.   The main approaches to the measurement of mental
load and effort will now be reviewed briefly, and the pros and
cons of these different techniques summarized.   No claims are
made for the definitive nature of this review.   As Wierwille &
Williges (1978) pointed out in their recent methodological assessment of 400 publications on this topic:   "...the literature on
workload is so diverse that categorization on the part of the
reader ... is almost intuitive".   However, it seems reasonable
to dichotomize the various approaches, and consider:   (a)
direct assessments of tasks, compared with,   (b) techniques for
the indirect assessment of load and effort.


## 2.   DIRECT TASK ASSESSMENT


### 2.1.   Subjective evaluation

   This is a widely used, rough-and-ready method of assessing
mental demands of existing or simulated tasks.   It clearly has
a fairly high face validity and it may often provide acceptable
comparative information on a limited number of alternative task
designs.   However, subjective assessments are usually performed
by people who are skilled in the tasks under consideration. This
means that many aspects of their performance will be pre-

programmed, or automatized (Bahrick et al. 1954), that is, carried out without complete conscious awareness of detailed behaviour. Thus, 'expert' opinions will not accurately reflect the mental effort which a task requires of individuals performing at different levels of skill. For example, using skilled drivers to assess the handling characteristics of vehicles may provide misleading information on the naive driver's ability to control those vehicles in an emergency.

## 2.2. Protocol analysis

This method requires operators to verbalize their performance of the task, either concurrently with actual task performance, or off the job. The results can be extremely informative on the mental demands of the job, especially of course for tasks in which many of the operator's responses are covert. One disadvantage of concurrent verbalization of performance is that during peak demands there may be insufficient time for the operator to provide complete information. Again, there is the possibility that actions normally performed below the level of conscious awareness will not be reported at all. Off the job, verbalization may simply omit certain actions which seem self-evident. The method is therefore somewhat unreliable, although it may be improved by the additional use of rating scale techniques to score the demands of sub-tasks. However, it is of course impossible to use this technique at the design stage of new tasks, unless high-grade simulation is available.

## 2.3. Task analysis

This is a more systematic method of investigating mental effort, and can be employed on existing tasks as well as for task design. It involves the breakdown of task performance from the operational level to the very detailed components of basic behaviour. The main advantage of the analytic approach is that it permits the task designer to distribute the operator's mental effort over time and thus avoid overload. It is also extremely useful in allocating demands to various members of a crew, where load has to be shared.

A comparable objective is also served by task synthesis, where overall mental load is computed from summation of sub-task demands.

It is appropriate to mention here one or two theoretical issues which have been important for the development of analytic and synthetic techniques. The first is the distinction made by Conrad (1955) between 'load stress' and 'speed stress'. In this connection, 'load' is a function of the number of sources of information employed on the task, and 'speed' is the rate at which information is transmitted from one or more of the sources. 'Speed stress', resulting from excessive rates of information transmission, is usually found to be the more critical parameter of task design. It can often be dealt with by allowing the individual to operate in a self-paced, rather than a forced-paced manner. For example, Knowles et al. (1953) found that forced pacing produced a 36% error rate on a task which was error-free under self pacing, even though the average rate of information presentation was comparable for both conditions. Chiles & Alluisi (1979) conclude from studies such as this that speed-of-response constitutes a valid and reliable means of specifying workload. One must of course take into account the fact that, at higher levels of workload, speed stress and load stress interact and both must therefore be considered in practical assessments of mental effort (Goldstein & Dorfman 1978).

A second theoretical development in this field followed the introduction of information theory to psychology (Quastler 1955). This offered the possibility that mental load and effort could be measured without regard to the specific nature of task demands and operator performance. Load would be measurable simply in 'bits' of information and speed in 'bits per second'. The investigation of choice reaction tasks was particularly aided by Hick (1952) who demonstrated that reaction time was a linear function of the amount of information transmitted by an individual. A number of models were developed which regarded man as operating like a single channel of limited capacity (e.g., Broadbent 1958). Thus, mental effort could be related to the probability that different sources of information would simultaneously demand attention.

This approach was developed by Senders (1955, 1970) for the assessment of the mental effort required in monitoring complex visual displays. It was thus possible to specify task demands in terms of the frequency with which visual fixations were required and the fixation duration necessary to extract information from individual sources. A technology of eye-movement recording seems to have developed largely to serve this method of investigating workload. Where computer simulations of displays were possible, the method was particularly useful in designing new systems which could be operated without undue speed or load stress.

One disadvantage of this approach was its unreliability in the
design of systems which employed information transmission via
different sensory modalities.   This problem has intensified in
recent years, as more and more evidence accumulates that man
can act as a multiprocessor of information, rather than as a
single channel (e.g. Gopher & Navon 1980, Hawkins et al. 1978,
McLeod 1977).   Thus, mental effort appears not to be a simple
function of simultaneous informational demands from a display.
Rather, it is a function of competition among a range of concurrent task demands for one or more of a set of limited processing resources.   Clearly, this intensifies the difficulty of quantifying mental effort, especially under conditions in which an
individual can switch strategies so as to reallocate the distribution
of demands on his various mental resources.   (See Welford 1978
for a more detailed discussion of strategy effects on workload).


2.4.   Performance measurement

For tasks which already exist, or which can be validly simulated,
mental effort can often be inferred from direct measurement of
speed and accuracy of performance.   Clearly, this method has
a high face validity.   It is especially useful for the investigation
of tasks where occasional errors are acceptable, since these can
provide a more comprehensive measure of work overload, under-
load and performance degradation.
   Probably the major theoretical development in this field, at
least for the study of monitoring-type tasks, was the introduction of signal detection theory to psychology (Green & Swets 1966).
This permitted even extremely small error rates of human observers to be treated mathematically in order to derive two independent indices of performance:   one reflecting the observer's
perceptual sensitivity, the other reflecting response bias, or
the operator's decision criterion.   Thus, effects of mental workload on covert decision processes could be investigated with much
greater precision than was previously possible, when small errors
of omission and commission were largely regarded as uninforma-
tive and negligible.   The method has its limitations, of course,
where mental effort has to be devoted concurrently to task
components which cannot be incorporated within the theory of
signal detection.
   A second theoretical development of relevance here was the
demonstration of an inverted-U-shaped relationship between
performance and 'arousal'; where 'arousal' is seen to be the
result of a complex interaction between task demands, environmental conditions and the individual performer's level of

physiological activation (e.g., Corcoran 1965). Both above
and below some optimal level of arousal, performance on a task
was shown to decline. Research on circadian rhythms in physiological activation shows that the mental effort required for
optimal task performance will vary over time-of-day and between
individuals (Blake 1967, Colquhoun 1971). Work such as this
highlights the absolute necessity to control for time-of-day and
individual differences in any attempt to infer mental effort directly
from performance measurement. Indeed, it might even be claimed
that such research brings into question the practical usefulness
of mental load and effort as basic concepts.

However, the theory has been useful in allowing us to understand that performance on certain tasks cannot always be
improved by the expenditure of greater effort (Kahneman 1973).
Norman & Bobrow (1975) have developed the theory further to
show that the discontinuity in the function relating effort to
performance depends upon whether individual information processing is data-limited or resource-limited.


3.   INDIRECT TASK ASSESSMENT

Given the practical difficulties in direct assessment of task
load and effort, especially where operators are highly motivated
to avoid error and thus where they provide few data on overt
task performance, researchers have turned to a variety of
indirect measurement techniques. Only the two main contenders
will be mentioned here.

3.1.   Dual task performance

This developed from the concept of man as an information
processing channel of limited capacity. Under normal conditions
of working, an operator would usually be performing within his
capacity and thus produce no error data which might be used to
assess mental effort. However, by combining the primary task
in question with another, performed concurrently, channel
capacity could be exceeded and measurable errors thus produced.
The basic model is illustrated in figure 2, taken from an early
paper by Brown (1964).

By instruction, errors could be concentrated within the secondary or the primary task, depending upon their acceptability
and whether one was interested in measuring 'reserve capacity',
as in figure 2, or in identifying primary task difficulty more
directly.

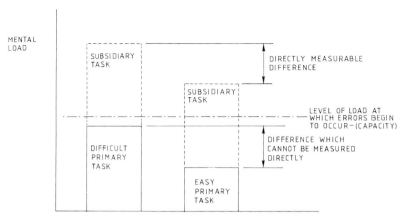

DIFFERING PRIMARY TASKS

FIGURE 2.   A model of the 'dual task' method of evaluating workload and mental effort, derived from early theories of man as a single information processing channel having limited capacity.

This method has been widely used in a variety of applications. For a comprehensive survey of the post-1965 literature on the subject, see Ogden et al. (1979), who were forced to conclude that it was impossible to identify a single 'best' task to use for indirect assessments of workload, or even specify a preferred category of secondary tasks.   Many years ago, Knowles (1963) attempted to list the desirable characteristics of secondary tasks, but some of these have since been disputed.   For example, Brown (1978) suggests that the method works best if the second task is performed subsidiary to the task being investigated and if it presents discrete stimuli of constant load, on a forced-pace schedule, competing for processing resources only.   This personal preference is based upon a reasoned argument that any other approach is likely to produce indices of mental effort which are at best unreliable, and at worst invalid.

3.2.   Physiological indices

A number of researchers have attempted to infer mental load and effort from the measurement of physiological responses to task demands and the individual's reactions to them (e.g., Wierwille 1979). Perhaps the more popular indices studied, largely because of the relative ease with which data can be collected in the laboratory and the field, are heart rate, heart rate variability, muscle tension and the Galvanic Skin Response. Respiration rate, pupillary dilation, EEG, cortical evoked potential, flicker fusion frequency and body-fluid analysis have also been used to infer effects of load and speed stress.

The method assumes that overload is accompanied by emotional stress, which can be measured and traced back to certain characteristics of the task in question. The attraction here, for psychologists, is that one avoids any assumptions about the single-channel nature of human information processing. In addition, many of the physiological responses are assumed to be involuntary and thus to provide a cleaner measure of workload effects than many behavioural measures, which are under the conscious control of the individual. Wierwille (1979) concludes that the more promising measures of workload and effort are pupil dilation, cortical evoked potentials and body-fluid analysis, while recognizing that much more research and development work is needed to refine the techniques used in measuring these indices.

There seem to be two main problems with this approach: (a) the equipment needed to collect data is often incompatible with real-life working conditions; and (b) the indices may be better indicators of the operator's perceived mental effort and of the perceived consequences of error than they are of actual task demands. In addition, of course, few data may be obtainable below some acceptable level of work stress, where the operator is performing effectively with 'zero risk' (e.g., Näätänen & Summala 1976).


4.   CONCLUSIONS

Most writers in this field agree that there is no single best way of assessing mental effort and that no single measure will, in any case, suffice completely. Most would probably agree with Wierwille and Willeges (1979) that any serious attempt to measure mental load and effort ought to combine several alternative techniques. An acceptable battery of tests would probably in-clude subjective assessment, task analysis, a dual task measure of reserve capacity and one or more appropriate physiological measures. Further research may allow us to refine this some-what cumbersome method of measuring mental effort, but we have to recognize that the multidimensional nature of mental load pro-hibits any easy solutions.

REFERENCES

Bahrick, H.P., Noble, M. & Fitts, P.M., 1954, Extra-task performance as a measure of learning a primary task. Journal of Experimental Psychology, 48, 298-302.
Bainbridge, E.A., 1974, Problems in the assessment of mental load. Le Travail Humain, 37, 279-302.
Blake, M.J.F., 1967, Relationship between circadian rhythm of body temperature and introversion-extraversion. Nature, 215, 896-897.
Broadbent, D.E., 1958, Perception and Communication (London: Pergamon Press).
Brown, I.D., 1964, The measurement of perceptual load and reserve capacity. Transactions of the Association of Industrial Medical Officers, 14, 44-49.
Brown, I.D., 1978, Dual task methods of assessing workload. Ergonomics, 21, 221-224.
Chiles, W.D & Alluisi, E.A., 1979, On the specification of operator or occupational workload with performance-measurement methods. Human Factors, 21, 515-528.
Colquhoun, W.P., 1971, Circadian variations in mental efficiency. In Biological Rhythms and Human Performance, edited by W.P. Colquhoun (London: Academic Press), p.39.
Conrad, R., 1955, Adaptation to time in a sensorimotor skill. Journal of Experimental Psychology, 49, 115-121.
Corcoran, D.W.J., 1965, Personality and the inverted-U-relationship. British Journal of Psychology, 56, 267-273.
Dornic, S., 1977, Mental load, effort, and individual differences. University of Stockholm, Department of Psychology Report No. 509.
Goldstein, I.L. & Dorfman, P.W., 1978, Speed and load stress as determinants of performance in a time sharing task. Human Factors, 20, 603-609.
Gopher, D. & Navon, D., 1980, How is performance limited: testing the notion of central capacity. Acta Psychologica, 46, 161-180.
Green, D.M. & Swets, J.A., 1966, Signal Detection Theory and Psychophysics (New York: John Wiley).
Hawkins, H.L., Church, M. & de Lamos, S., 1978, Time-sharing is not a unitary ability. University of Oregon, Center for Cognitive and Perceptual Research, Technical Report No.2.
Hick, W.E., 1952, On the rate of gain of information. Quarterly Journal of Experimental Psychology, 4, 11-26.
Kahneman, D., 1973, Attention and Effort (Englewood Cliffs, New Jersey: Prentice-Hall).
Knowles, W.B., 1963, Operator loading tasks. Human Factors, 5, 155-161.

Knowles, W.B., Garvey, W.D. & Newlin, E.P., 1953, The effect of speed and load on display-control relationships. Journal of Experimental Psychology, 46, 65-75.

McLeod, P.D., 1977, A dual-task response modality effect: support for multiprocessor models of attention. Quarterly Journal of Experimental Psychology, 29, 651-667.

Näätänen, R. & Summala, H., 1976, Road-User Behaviour and Traffic Accidents (Oxford: North-Holland).

Norman, D.A. & Bobrow, D.G., 1975, On data-limited and resource-limited processes. Cognitive Psychology, 7, 44-64.

Ogden, G.D., Levine, J.M. & Eisner, E.J., 1979, Measurement of workload by secondary tasks. Human Factors, 21, 529-548.

Quastler, H. (editor), 1955, Information Theory in Psychology (Glencoe, Illinois: Free Press).

Roitsch, P.A., Babcock, G.L. & Edmunds, W.W., 1977, Human Factors Report on the Tenerife Accident (Engineering and Air Safety, Washington, D.C.: Air Line Pilots Association).

Sell, R.G. & Shipley, P. (editors), 1979, Satisfactions in Work Design: Ergonomics and Other Approaches (London: Taylor & Francis Ltd).

Senders, J.W., 1955, Man's capacity to use information from complex displays. In Information Theory in Psychology, edited by H. Quastler (Glenco, Illinois: Free Press).

Senders, J.W., 1970, The estimation of operator workload in complex systems. In Systems Psychology, edited by K.B. De Greene (New York: McGraw-Hill).

Welford, A.T. (editor), 1974, Man Under Stress (London: Taylor & Francis Ltd).

Welford, A.T., 1978, Mental workload as a function of demand capacity, strategy and skill. Ergonomics, 21, 151-167.

Wierwille, W.W., 1979, Physiological measures of aircrew mental workload. Human Factors, 21, 575-593.

Wierwille, W.W. & Williges, R.C., 1978, Survey and analysis of operator workload assessment techniques (Blackburg, Virginia: Systemetrics Report S-78-101).

# DETERMINANTS OF CAPACITY FOR PHYSICAL WORK

J.E. Cotes, J.W. Reed and I.L. Mortimore

Respiration and Exercise Laboratory,
Departments of Occupational Health and Hygiene and Physiology,
University of Newcastle upon Tyne,
24 Claremont Place, Newcastle upon Tyne NE2 4AA

## 1. INTRODUCTION

Man, being a mammal, has an ability to generate a high power output relative to body weight under a wide range of environmental conditions. This paper describes attributes which contribute to the work capacity and the factors which influence them. These include inheritance, the processes of growth and decay, the level of customary activity and environmental factors, both physiological and pathological. Observations are also made on methodology.

## 2. EVIDENCE FOR THE DETERMINANTS OF EXERCISE CAPACITY

The evidence on determinants of exercise capacity comes from detailed cross-sectional studies of groups of subjects, and from longitudinal studies of changes during growth and ageing. The effects of training or detraining, heredity, and comparisons across ethnic groups have added to our knowledge. Unfortunately many such studies have been biased by the initial selection of subjects; this is partly the fault of the investigator, but in the case of exercise studies, self-selection by the subject, reflecting his or her willingness to undertake test exercise, is difficult to avoid. For mild exercise this factor may not be important, but where it is proposed to study strenuous or maximal exercise a proportion of potential subjects may be unable or unwilling to oblige. In these circumstances, the findings reflect the characteristics of athletic subjects and sportsmen and may not be representative of the general population. Thus, there is a need for socially acceptable submaximal tests which yield useful information about exercise capacity. In this context, the cardiac frequency measured during a conventional progressive exercise test has often been used to

TABLE 1. Types of effort

| | Explosive | Sustained |
|---|---|---|
| Examples | Leap, lift, throw, turn, sprint | Climb, run, saw, build |
| Main determinants | Muscle strength and skill | Capacity of $O_2$ transport chain (lungs to mitochondria) |
| Energy sources, primary | Adenosine triphosphate, Phosphocreatine | |
| Energy sources, secondary | Glycolysis, glycogenolysis | Oxidation of carbohydrate, fat and protein |
| Metabolic pathway | Phosphorylation of carbohydrate to pyruvate and energy | |
| | Anaerobic (few stages) oxidation of NADH (no energy release) ⟶ lactate | Aerobic (many stages) Krebs cycle, cytochrome chain + energy ⟶ $CO_2$ and $H_2O$ |
| Location in cell | Cytoplasm | Mitochondria |
| Predominant cell | 11 A and B (fast twitch) | I and II C (slow twitch) |
| Predominant enzymes | Lactate dehydrogenase, myokinase | Myosin ATPase, succinate dehydrogenase, etc. |
| Oxidative capacity | Low | High |
| Recovery | Slow | Rapid |

predict maximal effort. Because of the inherent errors in this method, alone it provides a poor guide; more information may be obtained by also studying components of the oxygen transport chain. Results may also be unknowingly biased by additional factors; to avoid this, allowance may need to be made for age, sex, physique, nutrition, smoking, alcohol, activity, climate, altitude, ethnic group, disease and other factors. Longitudinal exercise studies are liable to the additional error that lapses may occur predominantly among the less fit or frailer members of the initial sample, so their loss may bias the result. Genetic studies may suffer from any of these difficulties: in the case of twin studies, they are also liable to error on account of small numbers or a wide age range of twin pairs. For studies across ethnic groups, the extent of environmental differences may also be underestimated.

## 3. ANAEROBIC AND AEROBIC WORK

The energy for muscular contraction is provided initially by the breakdown of energy-rich compounds present in the muscle fibres. Subsequently, the process is completed by the removal of hydrogen, through combination with oxygen. The throughput of this aerobic metabolic pathway effectively determines the capacity for sustained work. By contrast, explosive effort of brief duration is performed initially without recourse to oxygen, though oxygen is needed for subsequent recovery. The principal features of these two types of effort are summarized in table 1.

The determinants of the capacity for explosive effort are principally muscle strength and skill; those of the capacity for sustained effort are the structural and functional dimensions and state of readiness for exercise of the oxygen transport chain. The links in the chain include lungs, blood, heart, systemic and pulmonary circulations, muscle circulation and oxidative enzymes present in muscle mitochondria. In normal healthy man exercising at sea-level, the limits of exercise are set by the cardiovascular system – at maximal levels of cardiac output, exercise ventilation is usually only 70–80% of maximum potential.

However, all components are usually matched for size, so subjects with a high capacity for exercise have relatively large dimensions for all of them. This is illustrated for some variables in figure 1. Correlations between work capacity, anaerobic threshold and the fibre type and oxidative capacity *in vitro* of samples of vastus lateralis muscle are given in table 2.

These dimensional components of the capacity for exercise

TABLE 2. Relationship of muscle cell type and respiratory capacity to lactate threshold and maximal oxygen uptake for 13 male subjects ages 14-34 years (Ivy et al. 1980).

| Index | | Range | Correlation coefficients | | | |
|---|---|---|---|---|---|---|
| | | | $Max.O_2$ | LT | FT | OC |
| Max. $O_2$ uptake (ml $kg^{-1}$ $min^{-1}$) | (max.$O_2$) | 38-65 | 1.0 | | | |
| Lactate threshold (ml $kg^{-1}$ $min^{-1}$) | (LT) | 15-38 | 0.91 | 1.0 | | |
| Fibre type I (area %) | (FT) | 24-80 | 0.67 | 0.73 | 1.0 | |
| Oxidative capacity (ml $g^{-1}$ $h^{-1}$) | (OC) | 1.3-3.9 | 0.83 | 0.94 | 0.70 | 1.0 |

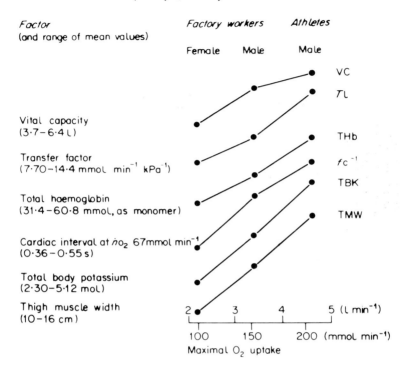

FIGURE 1. Relationship to oxygen uptake of some factors which
contribute to the capacity for exercise (from Cotes
and Davies 1969).

are themselves influenced by a number of factors of which some
are listed in Section (2) as possibly needing to be taken into
account when planning a study of analysing the results. A few
of the interactions are indicated in figure 2.

4.   INHERITANCE

     The overall body dimensions, the muscularity and the size
of the lungs relative to stature or trunk length are all influenced by genetic constitution (table 3), but except in a few
instances the proportion of variance explained by inheritance
is not known.   Much of the available evidence is derived from

TABLE 3. Summary of some factors which influence cardio-respiratory and muscle structure and function, including genetic constitution, childhood growth, physical activity and age.

| | Genetic factor | Growth | Activity | Ageing |
|---|---|---|---|---|
| **Cardiovascular system** | | | | |
| Heart, size | + | ← | ← | ← |
| stroke volume | | ← | ← | → |
| frequency submax. | | → | → | → |
|     max. exercise | | → | − | → |
| output   submax | | − | − | → |
|     max. exercise | | ← | ← | → |
| | | | | |
| Blood pressure, rest | + | ← | − | ← |
| submax. exercise | | | | ← |
| static | | | | ← |
| | | | | |
| **Blood** | | | | |
| Volume | | ← | ← | → |
| Haemoglobin concentration | | ← | →3 | → |
| Red-cell concentration | | ← | →3 | − |
| $O_2$ dissociation curve | + | ↓? | − | − |
| Total haemoglobin | | ← | ← | → |
| Arterial $O_2$ tension | | ← | − | → |
| Max. arterio-venous $O_2$ diff. | | ← | ← | → |
| | | | | |
| **Skeleto-muscular system** | | | | |
| Muscle strength | + | ← | ← | → |

| | | | | |
|---|---|---|---|---|
| Fibre diameter | + | ← | ← | → |
| length | | ← | − | − |
| number | + | − | ← | → |
| proportion type I | + | − | ← | − |
| oxidative capacity | + | ← | ← | − |
| Capillary density | | ← | ← | − |
| Bone diameter | + | ← | ← | − |
| length | + | ← | ←¹ | − |
| mineral content | | ← | ← | |
| Cartilage, thickness of | + | ← | ← | → |
| Respiratory system | | | | |
| Total lung capacity | + | ← | ← | − |
| Ventilatory capacity | | ← | ←¹ | → |
| Capacity to transfer gas | + | ← | ←¹ | → |
| Respiratory sensitivity - $CO_2$ | + | − | ←¹ | − |
|      - $O_2$ | + | → | − | |
| Vascular resistance | + | −² | − | ← |
| Exercise ventilation - submax. | | − | → | ← |
|      max. | | ← | ← | → |

[1]during period of growth, [2]after three months, [3]immediate effect

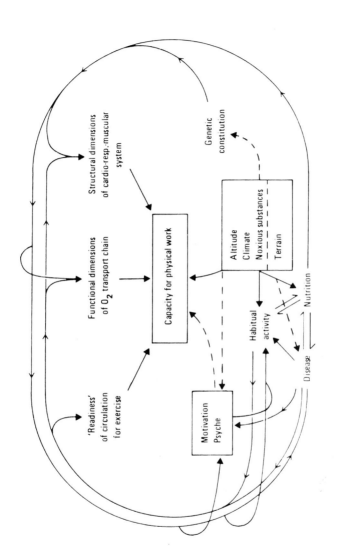

FIGURE 2. Determinants of capacity for physical work.

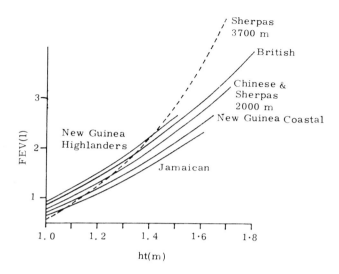

FIGURE 3. Forced expiratory volume (FEV$_{1.0}$) relative to stature in children of different ethnic groups (for sources see Cotes 1979).

animal studies, but lung function and exercise performance have mainly been studied in man. The findings should be viewed cautiously for the reasons given in Section 2.

The inheritance of 'muscularity' extends to muscle strength and size, including the diameter and number of muscle fibres, the proportion of type I fibres and the oxidative capacity, but many of these indices are intercorrelated. There is a similar genetic component to the skeletal dimensions and to the volume and stroke output of the heart. The size of the lung and hence the ventilatory capacity differs between ethnic groups, with people of European descent having larger lungs relative to stature compared with most other ethnic groups. The changes are most readily illustrated for children (figure 3) as among them there is little or no component due to age, which in adults is an additional complicating factor. However, after allowing for the effects of smoking there is no evidence that the lungs of people of different ethnic groups age at different rates. The respiratory sensitivity to oxygen and carbon dioxide, and the sensitivity of the pulmonary vascular resistance to hypoxia, also show pronounced differences between individuals which are at least partly genetically determined; in the case of the former indices, allowance needs to be made for the size of the lungs. Within ethnic groups, the sensitivity is correlated with other attributes which reflect the capacity for exercise, but this is not the case between ethnic groups. However, while ethnic group influences some components of respiratory function, it does not have a proportionate effect upon the performance during exericse. This is probably also the case for the difference between the sexes. The lower exercise capacity of women compared with men is mostly due to the smaller functional dimensions of their oxygen transport chain and to their having less muscle. Both these features are to some extent genetically determined, but the association is at least partly indirect due to earlier cessation of growth in women and to behavioural differences which are partly culturally determined; the direct genetic component is probably small.

5. GROWTH DURING CHILDHOOD AND ADOLESCENCE

Almost all organs and tissues enlarge during growth, and this process is accompanied by an increase in functional capacity of components of the oxygen transport chain (table 3); the ability to undertake exercise increases in consequence. The change is most conspicuous in the case of cardiac frequency which, during submaximal exercise of constant intensity, falls as the child grows; this is due to an increase in the stroke output of the heart. The change is correlated with an increase in quantity of body muscle. Similarly, the respiratory frequency falls due to growth of the lung, leading to an increase in the maximal tidal volume. However, the relationship of cardiac output to uptake of oxygen and, except at high rates of work, of ventilation to uptake of oxygen, are relatively constant and independent of body size over the age range 7-25 years. Thus, during submaximal exercise, the mass flow of oxygen is largely independent of the dimensions of the system. The exception reflects the smaller absolute threshold for anaerobic work of children compared with adults; this leads to the ventilation being stimulated by lactic acid, and hence to hyperventilation, at intensities of effort which would be sustained by aerobic metabolism in older subjects (table 1). Thus, the functional changes are mainly a simple consequence of growth, but this does not occur uniformly. For example, in boys during adolescence the torso grows relative to the legs, and the lungs grow relative to the stature and sitting height. The additional increment of lung function is related to age up to about 23 years, when the lung function is at its peak and is poised for

the subsequent slow decline discussed in Section 7. It is not
known if the improvement is related to an increase in muscle
strength. In females there is no comparable improvement in
lung function during the teenage period, but the decline with
age is rather less. Evidence on the magnitude of other components of the oxygen transport chain relative to body size
is incomplete, but a spurt during adolescence may well occur.

## 6. HABITUAL ACTIVITY

The level of habitual activity influences both the structural
and the functional components of the oxygen transport system.
The interdependence is probably greatest during the period of
growth, when the whole system appears to share in the adaptation. However, there is need for caution in interpretation, as
some of the evidence is biased by the selective factors discussed earlier.
Relatively brief exercise of 10 min duration which is
repeated as infrequently as every third day can have a pronounced effect, but longer, more frequent exercise provides a
greater stimulus. Submaximal exercise which is of sufficient
intensity to materially increase the exercise cardiac frequency
affects the active muscles themselves; exercise of large muscle
groups also exerts a training effect upon the heart. The
changes in muscle structure include the size and type of the
fibres, the density of capillaries, the number of mitochondria,
the content of enzymes and the oxidative capacity (table 3).
These changes lead to the trained muscle requiring less blood
flow for a given rate of submaximal energy expenditure than
does an untrained muscle; the decreased sympathetic outflow
affects the circulatory response to the exercise, including the
rise in cardiac frequency and the extent of diversion of blood
flow from the viscera. At the same time, the enhancement of
the muscle's oxidative capacity increases its capacity for sustained work.
As stated previously, exercise of large muscle groups has
a training effect upon the heart itself, such that myocardial
contractility is improved; this serves to increase the stroke
volume. The change is enhanced by an increase in quantity
of heart muscle if the training is of sufficient duration. Hence
the maximal cardiac output and the maximal muscle blood flow
are increased, and these changes further improve the capacity
for exercise. Some circulatory effects of training are reviewed
by Clausen (1977) and are summarized in table 4. They include
an acceleration in the circulatory adaption to exercise; this mini-
mizes the oxygen deficit in muscle at the start of the exercise

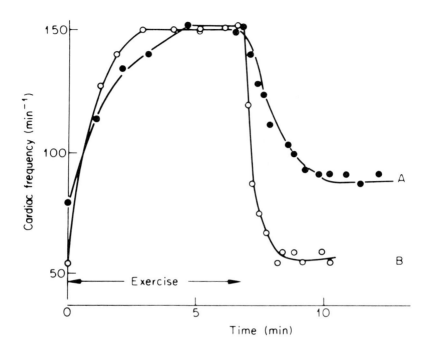

FIGURE 4. Acceleration and decline of cardiac frequency at
         start and end of exercise of comparable severity in
         a world-class athlete (B) and a subject having a low
         capacity for exercise (A, from Cotes 1979).

and the recovery occurs sooner (figure 4). Other effects of a
high level of habitual activity are summarized in table 3. Some
occur consistently, provided the stimulus is of sufficient intensity. Others, including the changes in lung function, only
occur in some circumstances. Thus, the vital capacity enlarges
as a consequence of work with the arms and back, and is positively correlated with the aerobic capacity. Evidence for a
causative association comes from a study of Chinese children in
Hong Kong who were born to lifestyles of high and low activity
by a sequence of events which was outside their control and
was probably independent of the vital capacities of their parents.
In this study, the active children had large vital capacities

TABLE 4. Circulatory effects of physical training
(after Clausen 1977).

| | Rest | Submaximal exercise | | Maximal exercise |
| | | Primary effects | Secondary effects | |
| --- | --- | --- | --- | --- |
| Muscle blood flow | | ↓ * | | ↑ * |
| $O_2$ extraction from blood | | | ↑ * | |
| Visceral blood flow | | ↑ * | | |
| Cardiac frequency | ↓ | ↓ * | | – |
| Cardiac stroke volume | ↑ | | ↑ * | ↑ |
| Cardiac output | – | – | | ↑ |
| Oxygen uptake | – | – | | ↑† |

*Effects which are probably entirely due to changes in the trained muscles, including an increase in oxidative capacity and reduction in sympathetic outflow on submaximal exercise.

†Effect is mainly due to changes in the trained muscle.

relative to body size; their ability to ventilate the lung ($FEV_{1.0}$) was also increased. The picture is similar for the transfer factor, or diffusing capacity of the lung for carbon monoxide (T𝑙) in that athletic subjects have high values at rest and during submaximal exercise. It should be noted, however, that physical training in adult life only increases the value obtained during maximal exercise. The transfer factor at rest when standardized for age, body size and haemoglobin concentration is correlated with indices which reflect the aerobic capacity (figure 5). Increased rates of increase of T𝑙 with growth have been reported in child swimmers, but this may have been a consequence of high initial values. More probably, it is part of a general adaptation whereby physical activity during growth enhances the development of the oxygen transport system and hence the intercorrelations between its components. A contributing factor to this process may be the release of growth hormone which occurs during exercise; in sufficient quantity, this hormone increases the dimensions and hence the function of the lung.

7. AGEING

The processes of growth and development which accelerate

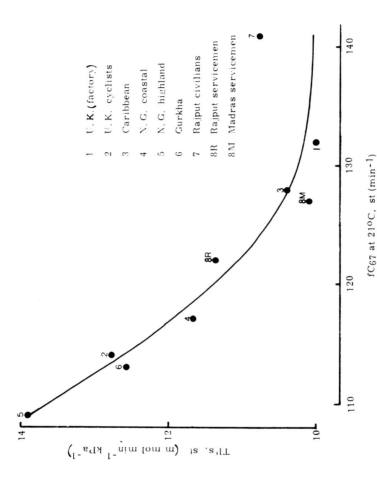

FIGURE 5. Relationship of transfer factor standardized for age, stature, oxygen tension and haemoglobin concentration (T's. st) to exercise cardiac frequency standardized for oxygen uptake, fat-free mass and ambient temperature (fc₆₇) for groups of healthy young adult males.

at puberty tail off thereafter to reach a plateau which in males
is attained at about the 25th year. But it is short-lived.
Already at this time some aspects of function are beginning to
decline, though others may be capable of further development
as a result of physical training. This is the case for marathon
runners who, if they are dedicated and persistent, attain a
peak of performance in their early 30's. However, by this
time the function of the lung, including the ventilatory capacity
and the ability to transfer gas, have on average declined from
the peak by about 7%. The deterioration is greater in smokers
than in non-smokers, though not in proportion to the amount
smoked. However, for reasons which are probably related to
self-selection, the smokers on average have superior lung function to start with. The changes are associated with loss of lung
elastic recoil pressure and of alveoli, and have the effect that
if they were to progress linearly the function of the lung would
fall to zero by about the 120th year. In practice, the linear
model has been found to apply up to the 90th year, so the inference is not unreasonable! There is a similar overall decline in
maximal cardiac output and work capacity, but evidence on the
pattern of decline is less complete; this is mainly because of
the difficulty of conducting epidemiological studies of maximal
exercise. For example, in a recent investigation of asymptoma-
tic middle-aged men who were performing physically active jobs
in heavy industry, we observed electrocardiographic abnormali-
ties which were sufficient to contraindicate the maximal exercise
test in 21% of subjects; 36% were unable to achieve an acceptable
maximal level of exercise during the test and in only 43% was a
realistic maximal performance observed. The decline in work
capacity has both central and peripheral components. The former include reductions in stroke volume and in submaximal and
maximal cardiac frequency; thus, the useful work of the heart
is reduced. However, the total work is not reduced to the
same extent as there is an increase in the systemic blood pressure and vascular resistance. The increase is apparent at rest
and during static effort as well as on dynamic exercise, and is
one factor contributing to an increase in the size of the heart.

The central component of the decline in work capacity is
related in part to a reduction in sympathetic nervous activity.
The peripheral component is mainly a consequence of loss of
muscle substance, the fibres being both smaller and less numerous than in younger subjects. The proportion of type I fibres
and the oxidative capacity of the tissue is apparently only
affected to the extent that men and women reduce their habitual
physical activity as they get older. These aspects are summarized in table 3. The deterioration may be aggravated by an
increase in body mass with age and by the decline in lung

TABLE 5.　Effects of an altered thermal environment.

| | | Cold | Hot |
|---|---|---|---|
| Lungs | Total lung capacity | ↑[1,2] | - |
| | Ventilatory capacity | ↓[1,2] | - |
| | Capacity to transfer gas | ↑ | ↓ |
| | Respiratory sensitivity - $CO_2$ | ↓ | ↑ |
| | - $O_2$ | | |
| | Vascular resistance | | ↓ |
| | Exercise ventilation (relative to $nO_2$) | ↓ or ↑ | ↑ |
| Heart | Stroke volume | ↑ | ↓[1] |
| | Frequency - submaximal exercise | − | ↑ |
| | maximal | ↓[1] | − |
| Blood | Volume | | ↑[2] |
| | Skin blood flow | ↓[1] | ↑[1] |
| | Haemoglobin concentration | | ↓[2] |
| | Systemic pressure | ↑ | ↓ |
| | Arterial $O_2$ tension | ↓ | |
| Sweat rate | | | ↑[2] |
| Muscle | Oxidative capacity | ↓ | |
| | Basal metabolic rate and $O_2$ cost of activities | ↑ | ↓[2] |
| Skeleton | Joint movement | ↓ | − |

[1]Principal changes

[2]Long-term changes

function. Both processes contribute to the ventilatory cost of activities being increased. The latter also reduce the ventilatory ability. The degree of breathlessness experienced during performance of standard tasks is increased in consequence.

## 8. EFFECTS OF AN ALTERED THERMAL ENVIRONMENT

### 8.1. Heat

A high environmental temperature is compensated for by increased heat loss through the skin, mainly by evaporation of sweat. The process entails a redistribution of blood volume and blood flow which has the immediate effect of reducing the stroke volume of the heart and hence the capacity for exercise. This effect is enhanced by dehydration if the fluid lost in perspiration is not replaced. The peripheral vasodilation also leads to less blood being present in the lungs, and this reduces the capacity of the lung to transfer oxygen into the blood; however, the magnitude of the effect is small. Of greater importance is the enhanced rise of body temperature during exercise, which stimulates breathing both by a direct effect on the respiratory control centres in the brain and indirectly via an increase in respiratory sensitivity to carbon dioxide. The ventilation minute volume is increased in consequence. These and other changes are summarized in table 5.

Acclimatization to heat is mainly through an increase in sweat rate, but there is also an increase in blood volume, usually without a corresponding increase in total body haemoglobin, so the haemoglobin concentration falls slightly; however, the overall effect is beneficial. There is also a reduction in metabolic rate, which has the effect of reducing the oxygen cost of activities. Where the heat load is due to solar radiation, additional protection is provided by appropriate clothing and adoption of a suitable lifestyle. However, in most circumstances the adaptations are insufficient to maintain the capacity for exercise at the optimal level. The shortfall is greatest when the environment is adverse in other respects, including endemic diseases and deficiency of essential nutrients such as iron and salt.

### 8.2. Cold

A cold environment is particularly hostile, because of the risk of hypothermia; the physiological responses are of limited effectiveness and only slightly amenable to conditioning. Man's survival has depended on his ability to modify the environment in which he has existed, not to combat it. In addition, exposure to cold is disagreeable if not painful, and impairs the effectiveness of the operator as well as the subject. Thus, it is not surprising that the response to cold is less well documented than that to the other principal environmental hazards. Some of the effects are summarized in table 5.

The immediate response is constriction of superficial skin blood vessels and opening of arterio-venous anastomoses in the deep subcutaneous tissue. This change maximizes the insulating properties of the subcutaneous fat and is most effective in those with a high proportion of adipose tissue. However, the adaptation provides no protection for the extremities of the body, including ears, nose, fingers and toes, which are particularly vulnerable to injury from the cold. In the case of the fingers, loss of functions such as sensation and dexterity further reduces the ability to respond.

The peripheral vasoconstriction diverts blood not only into the central body core but also to the alveolar capillaries of the lungs; this enhances the ability to transfer oxygen into the blood, but also leads to overperfusion of some areas relative to their ventilation. The arterial oxygen tension diminishes in consequence. At the same time, the blood pressure and the stroke volume of the heart both increase but the maximal cardiac frequency falls. The hypoxaemia exerts a stimulant effect upon respiration via the central chemoreceptors, but this may be offset by depression of the respiratory centres by hypothermia. The hypothermia also reduces the oxidative capacity of muscles.

Active heat production is achieved by shivering and in other ways, including activation of brown fat. In the longer term, an increased production of thyroxine further raises the basal metabolic rate. Additional heat may be generated by voluntary effort (including those activities which are socially useful in a cold environment, such as cutting wood and shovelling snow). Provided the conditions are not too severe, the overall effect is favourable for the able-bodied section of the community who experience the benefits of a high level of physical activity described in Section 6. Their capacity for exercise is increased in consequence. However, in very cold environments, even with adequate thermal insulation and internal heat production, exposure to cold air may damage the lungs, and a recent report attributes impaired lung function, including an increase in total lung capacity and reduction in ability to ventilate the lungs, to this cause. Overall, cold increases the cost of submaximal exercise and reduces the maximal working capacity.

## 9. EFFECTS OF A CHANGE IN BAROMETRIC PRESSURE

### 9.1. High altitude

Acute ascent to high altitude or simulated altitude in a decompression chamber reduces the barometric pressure; this lowers the density of the respired gas and reduces the partial pressure of oxygen both in the atmosphere and in the arterial blood. The latter change sets in motion compensating adjustments, both immediate and long-term. In addition, if the hypoxaemia is material and especially if it is of sudden onset there may also be adverse consequences of which some are listed in table 6. If the ascent is on land, the situation may be complicated by the effects of cold and sometimes of dehydration or, in the case of high-altitude dwellers, of dietary deficiency, for example of iodine. There may also be an effect due to a high level of habitual activity, in that mountain dwellers tend to have a more physically demanding lifestyle.

Hypoxaemia increases the drive to respiration from the central chemoreceptors; the resulting hyperventilation eliminates additional carbon dioxide from the lungs and sets in motion compensating adjustments to the acid/base balance of the body; this is the first phase of altitude acclimatization. The hypoxaemia also leads to tachycardia. The hyperventilation and tachycardia are conspicuous during exercise; they have the effect of raising the alveolar oxygen tension to the greatest practical extent, and of increasing the cardiac output relative to the rate of work. These changes increase the delivery of oxygen to the tissues. At the same time, the reduction in density of the respired gas reduces both the work of breathing and the air resistance (drag) experienced during sprinting and cycling. During sojourn at altitude, the delivery and utilization of oxygen are enhanced by other changes. These include an increase in the haemoglobin concentration of blood, which is secondary to a greater production of the hormone erythropoietin. The oxygen dissociation curve is shifted to the right by a rise in the plasma concentration of 2,3-diphosphoglycerate, and the number of capillaries and the concentration of oxidative enzymes in the active muscles are increased. However, these changes do not fully compensate for the content of oxygen in blood being reduced by the hypoxaemia. Instead, anaerobic respiration occurs at a relatively low level of metabolism; this further stimulates breathing, and the resulting breathlessness also contributes to the exercise limitation. The changes are most marked before acclimatization on first exposure to altitude or on return to altitude from sea level, when the subject is also vulnerable to mountain sickness. The risk of mountain sickness

TABLE 6. Effects of acute and chronic exposure to changes in barometric pressure.

| | | Reduced | Increased |
|---|---|---|---|
| Lungs | Total lung capacity | ↑[2] | ↑[2] |
| | Ventilatory capacity | ↑ | ↓[1] |
| | Capacity to transfer gas | ↑[2] | – |
| | Respiratory sensitivity - $CO_2$ | ↑ | ↓ |
| |                   - $O_2$ | ↓ | |
| | Vascular resistance | ↑ | |
| | Exercise ventilation (submax. and max.) | ↑ | ↓ |
| Heart | Stroke volume | ↓ | |
| | Frequency - submaximal exercise | ↑ | – |
| |            maximal | ↓ | – |
| Blood | Volume | ↓ | |
| | Haemoglobin concentration | ↑[2] | – |
| | 2, 3-Diphosphoglycerate concentration | ↑[2] | |
| | Arterial $O_2$ tension | ↓[1] | |
| Muscle | Oxidative capacity | ↑[2] | |
| Nervous System | Function of | ↓ | ↓[1] |

[1]Principal changes

[2]Long term changes

may be reduced by prior treatment with diamox or frusamide, which respectively inhibit carbonic anhydrase and promote the elimination by the kidneys of retained fluid. The hypoxaemia also affects exercise capacity directly by raising the pulmonary vascular resistance, especially in susceptible individuals (Section 4), and, in some subjects, by depressing the maximal cardiac frequency. It may affect the capacity indirectly by impairing the quality of sleep.

In subjects whose exposure to altitude starts in childhood, further changes may occur. They include a reduction in respiratory sensitivity to hypoxaemia, enlargement of the lungs and an increase in the ability to transfer gas into the blood. These changes appear to be a consequence of the environment, but there may be a genetic component due to selection of subjects with appropriate characteristics. The extent to which the environmental factor is altitude or activity is not always apparent: in the case of young children in New Guinea, the possession of a large lung relative to stature appears to be related to altitude, but in the Himalayas the lung size of young children is independent of altitude over a wide range. Results of cross-sectional studies suggest that the increase with altitude develops during adolescence, but this is one of many topics on which there is need for more research.

The adverse effects of altitude upon exercise performance may be offset by the administration of oxygen during the exercise and preferably also during sleep, which is otherwise often disturbed, its restorative quality being impaired by periods of apnoea (Cheyne-Stokes breathing). Performance during sedentary tasks, including aviation, may be maintained by enclosing the subject in a pressurized cabin or by enrichment of the air with oxygen.

## 9.2. Increased barometric pressure

The barometric pressure is reduced at altitude on account of rarefaction of air; it is correspondingly increased in deep mines by a similar proportional amount, but with present mining practice the biological effects are small. Greater pressures are experienced beneath water during diving and also during work in caissons. Similar conditions may be reproduced in a compression chamber. The principal problems at depth are ergonomic ones, but at very high pressures the narcotic properties of gases, including nitrogen, are important. The physiological effects (table 6) are due to the altered physical properties of the gaseous environment, especially the increase in gas density. This increases the work of breathing and has the effect of reducing both the ventilatory capacity and the ventilation during exercise. The capacity for exercise is reduced in consequence. However, when the subjects are breathing compressed air, the impairment due to a rise in gas density is partially offset by an associated increase in the partial pressure of oxygen. This reduces the normal chemoreceptor drive to respiration and increases the content of oxygen in blood. Because of these changes, the deficit in exercise performance is less than would otherwise be the case.

The increase in respiratory work, together with the physical requirement for activity under water, exert a training effect upon the muscles of respiration. This may contribute to enlargement of the vital capacity, which is often already of above average size because of the selection for diving of physically active subjects. The enlargement may not extend to the lung airways, nor lead to improved ventilatory capacity - indeed, recent work suggests that in some subjects the capacity for exercise may be reduced in consequence. However, the evidence is inconclusive and more work is in progress.


## 10. OTHER CONDITIONS

### 10.1.   Iron deficiency

A diminution in intake of iron or increase in blood loss from any cause reduces the total amount of haemoglobin and enzymes containing iron present in the body. The latter include the oxidative enzymes in muscle, but the deficiency is usually slight. The anaemia reduces the oxygen capacity for blood, and this inevitably affects the capacity for exercise. However, even with chronic anaemia the impairment is relatively small and indices of performance during submaximal exercise are often normal. This is due to the compensating effects of an increase in blood volume, which raises the stroke volume of the heart, and a reduction in mixed venous oxygen tension. The ability of the lung to transfer oxygen into the blood is also reduced, but this is a second-order effect. During maximal exercise, the performance is reduced; the impairment is less than would be expected from the haemoglobin concentration, and probably reflects more the total body content of haemoglobin. The relationship of these variables might be expected to be similar to that for healthy subjects (figure 1).

TABLE 7. Effects of anaemia, carbon monoxide and ethanol

|  |  | Anaemia | Carbon monoxide | Ethanol |
|---|---|---|---|---|
| Lungs | Total lung capacity | − | − | − |
|  | Ventilatory capacity | − | − | − |
|  | Capacity to transfer gas | ↓ | ↓ | ↓ [2] |
|  | Respiratory sensitivity - $CO_2$ |  |  |  |
|  | $- O_2$ |  |  |  |
|  | Vascular resistance | ↓ | − | ↑ |
|  | Exercise ventilation | − | − |  |
| Heart | Stroke volume | ↑ | − | − |
|  | Frequency - submax. exercise | − |  | ↓ |
|  | - maximal | − | − | − |
| Blood | Volume | ↑ | − | ↑ |
|  | Haemoglobin concentration | ↓ [1] | − | ↓ |
|  | Arterial $O_2$ tension | − | − | ↓ [2] |
|  | A-V max. $O_2$ difference | ↓ | ↓ | − |
| Muscle | Oxidative capacity | ↓ | ↓ [1] | ↓ |
| Central nervous system | Function of | − | ↓ | ↓ |

[1] Principal changes

[2] Long-term changes

## 10.2.  Carbon monoxide poisoning

Inhalation of carbon monoxide from tobacco smoke or exposure to carbon monoxide gas reduces the ability of the blood to
transport oxygen.  This is due to both selective inactivation of
part of the haemoglobin molecule, which changes the shape of
the oxygen dissociation curve, and to inactivation of muscle
enzymes, which leads to muscle weakness.  Treatment is by
withdrawal from exposure and, for acute exposure, by adminis-
tering oxygen.

## 10.3.  Ethanol

The ingestion of ethanol leads to peripheral vasodilation
and an increase in cardiac frequency at rest and during submaximal exercise; however, the maximal frequency and stroke
volume are relatively unaffected, so the capacity for exercise
is not greatly reduced on this account.  Performance is also
affected by the cerebral and muscular effects of the drug,
which together contribute to impairment in proportion to its
blood concentration.  With prolonged exposure other changes
occur which further reduce the capacity for exercise, including
metabolic changes secondary to cirrhosis of the liver and, in
some cases, defective pulmonary gas transfer and hypoxaemia
secondary to changes in the lungs.

## 11. COMMENT

Much is now known about the immediate determinants of the
capacity for exercise and the ways in which these are influenced
by adverse environmental and other factors.  However, the
information has mostly been obtained on dedicated subjects.
Relatively little is known about the levels of performance of
whole populations.  Some reasons for this are given in Section
2.  The main difficulty has been in obtaining reasonably accurate estimates of capacity by methods which do not themselves
entail maximal exercise.  However, the position is now improving in this respect (table 2) and there is a prospect that before long more will be known about the long-term determinants
of exercise capacity in communities and their relevance for
health and wellbeing.

## 12. ACKNOWLEDGEMENTS

We are indebted to the Editor, Journal of Royal Society of Medicine for permission to publish figure 1, and to Messrs. Blackwell for permission to reproduce, from Lung Function: Assessment and Application in Medicine, figure 4. Figure 5 is adapted from one published by Messrs. Karger in Advances in Exercise Physiology.

## REFERENCES AND FURTHER READING

Astrand, P.-O. & Rodahl, K., 1977, Textbook of Work Physiology, 2nd edition (London: McGraw-Hill).

Blimkie, C.J.R., Cunningham, D.A. & Nichol, P.M., 1980, Gas transport capacity and echocardiographically determined cardiac size in children. Journal of Applied Physiology: Respiratory, Environmental and Exercise Physiology, 49, 994-999.

Blomqvist, G., Saltin, B. & Mitchell, J.H., 1970, Acute effects of ethanol ingestion on the response to submaximal and maximal exercise in man. Circulation, 42, 463-470.

Cerretelli, P., 1976, Limiting factors to oxygen transport on Mount Everest. Journal of Applied Physiology, 40, 658-667.

Clausen, J.P., 1977, Effect of physical training on cardiovascular adjustments to exercise in man. Physiological Reviews, 57, 779-815.

Cotes, J.E., 1979, Lung Function: Assessment and Application in Medicine, 4th edition (Oxford: Blackwell).

Cotes, J.E. & Davies, C.T.M., 1969, Factors underlying the capacity for exercise: a study in physiological anthropometry. Proceedings of Royal Society of Medicine, 62, 620-624.

Crosbie, W.A., Reed, J.W. & Clarke, M.C., 1979, Functional characteristics of the large lungs found in commercial divers. Journal of Applied Physiology: Respiratory, Environmental and Exercise Physiology, 46, 639-645.

Doell, D., Zutter, M. & Anthauisen, N.R., 1973, Ventilatory responses to hypercapnia and hypoxia at 1 and 4 ATA. Respiration Physiology, 18, 338-346.

Edholm, O.G. & Weiner, J.S., 1981, Principles and Practice of Human Physiology (London: Academic Press).

Editorial, 1980, Alcoholic heart disease. Lancet, 9, 961-962.

Heath, D. & Williams, D.R., 1977, Man at High Altitude (Edinburgh: Churchill Livingstone).

Ivy, J.L., Withers, R.T., Van Handel, P.J., Elger, D.H. & Costill, D.L., 1980, Muscle respiratory capacity and fibre type as determinants of the lactate threshold. Journal of

Applied Physiology: Respiratory, Environmental and Exercise Physiology, 48, 523-528.

Jokl, E., Anand, R.L. & Stoboy, H. (editors), 1976, Advances in Exercise Physiology (Basel: S. Karger).

Jones, P.R.M., Baker, F.M., Heywood, C. & Cotes, J.E., 1977, Ventilatory capacity in healthy Chinese children: relation to habitual activity. Annals of Human Biology, 4, 155-161.

Kronenberg, R.S. & Drage, C.W., 1973, Attenuation of the ventilatory and heart rate responses to hypoxia and hypercapnia with ageing in normal men. Journal of Clinical Investigation, 52, 1812-1819.

Kung, M., Tachmes, L., Birch, S.J., Fernandex, R.J., Abraham, W.M. & Sackner, M.A., 1980, Haemodynamics at rest and during exercise in comfortable, hot and cold environments. Measurement with a rebreathing technique. Bulletin Europeen de Physiopathologie Respiratoire, 16, 429-441.

Margaria, R., 1966, Exercise at Altitude (Amsterdam: Excerpta Medica Foundation).

Milvy, P. (editor) 1977, The Marathon: Physiological, medical, epidemiological and psychological studies. Annals of New York Academy of Sciences, 301, 1-1090.

Rowell, L.B., 1974, Human cardiovascular adjustments to exercise and thermal stress. Physiological Reviews, 54, 75-159.

Shephard, R.J., 1978, Human Physiological Work Capacity. (IBP 15) (Cambridge University Press).

Weiner, J.S. (editor) 1977, Physiological Variation and its Genetic Basis (London: Taylor & Francis).

# ENERGY EXPENDITURE, PRODUCTIVITY AND ENDEMIC DISEASE

K.J. Collins

Medical Research Council, London School of Hygiene and Tropical Medicine, Keppel Street, London WC1

## 1. INTRODUCTION

It has become an accepted axiom that the health of a nation profoundly affects its social and economic development. Nowhere is this more evident than in the tropical and sub-tropical regions of the world, where endemic tropical diseases play a major role in shaping hard-won achievements in development (Weisbrod 1961, Abel-Smith 1972, WHO 1973, El Batawi 1975). The debilitating effects of endemic disease on a population's work capacity and physical fitness has received scant study, and yet in many communities the capacity to perform physical work remains an important component of productivity. Undoubtedly, there exists a link between the prevalence and severity of disease, actual energy expenditure and productivity which should be susceptible to investigation and quantification. The reasons why few such studies have been attempted can be readily appreciated, for many developmental planners continue to regard some or all of these elements as non-quantifiable, and the most elementary attempts at investigating the influence of disease on working capacity and productivity disclose major shortcomings in relating experimental design to the actual working situation.

In order to measure the impact of any one major endemic disease there are a number of specific questions which need to be resolved. What, for example, is the respective reduction in work output contributed by acute illness as compared with long-term deterioration in performance associated with the chronic ill-effects? Is it possible to relate the severity of pathological effects, i.e., morbidity, to measurable physiological changes in work capacity? And where in tropical and sub-tropical regions do we encounter endemic disease which is not complicated by poly-parasitism? Those who are incapacitated, for however short a time, are automatically excluded from a population study of work performance, and there are many similar opportunities for sampling bias in the selection of test groups.

*65*

There are two essential components of energy-expenditure measurement. One is the potential or maximal work which can be performed under controlled (laboratory) conditions and the other is the actual work performance in the field. The question is whether either of these measurements can be translated into habitual energy expenditure which would relate to normal occupational activity. In addition, there are important socio-economic (Shephard 1980) as well as motivational and training/detraining determinants of working capacity which are difficult to assess. A variety of methods are available for measuring the physiological components of physical work. The energy cost of particular activities in the field are conventionally made by measurement of oxygen uptake during habitual physical activity, but there are many constraints on this type of field observation, as indicated above. One method favoured in our own investigations is to determine potential physical working capacity by ergometric tests of maximum aerobic capacity ($\dot{V}O_2$ max). This requires laboratory-based tests and an artificial work situation, but $\dot{V}O_2$ max is still regarded as a standard index of overall fitness or work capacity, subject to important reservations (Shephard 1980, Harrison et al. 1980). Obviously, the direct determination of maximal aerobic power involves a degree of forced exercise usually only easily attainable by the trained athlete. The alternative for less-fit subjects is to predict maximum oxygen intake from physiological responses to submaximum effort (Åstrand & Ryhming 1954, Margaria et al. 1965). It is well known that submaximal predictive methods are open to substantial individual errors and are dependent on factors such as age, habituation etc., but if confined to population studies there is less ground for criticism.

Finally, we have to consider productivity, a tangible but most difficult element to quantify in field investigations. In developing countries where continuous employment may not be the norm, and where the method of work is uncontrolled, productivity cannot readily be assessed with any degree of accuracy. At best, an estimate may be obtained from overall output figures or by records of absenteeism (Fenwick & Figenschou 1972, Davies 1973). Even if a standardized work test can be devised, there is the danger that the presence of observers and, in consequence, increased incentive will create an abnormal working situation.

In view of the many uncertainties inherent in investigations of this type, it may be asked why we even attempt to try to unravel these complex interrelationships. The motivation for our own studies has arisen from three main requirements. In the first place, some of the most fundamental facts about the effects of major endemic diseases on physiological function, particularly during physical work, are not known. Secondly, the potential for improving many aspects of community health in developing countries is great, and thirdly, in economic terms for the countries concerned, the stakes are high. One of the principal dilemmas facing administrators of aid programmes to industrially developing nations is to decide whether aid should be channelled directly into industrial resources or into health care in order to obtain the optimum use of resources. Few governments in developing countries are willing to devote finances which are in short supply to the expensive eradication of endemic disease, unless it can be conclusively shown that the disease has a measurable impact on the country's economy. The investigations described in this paper focus on the effects of one disease, schistosomiasis, which is estimated to infect more than 200 million people in the tropical and sub-tropical parts of the world. Physiological tests of work performance and measurements of field productivity have been made in populations in the Sudan where Schistosoma mansoni is endemic, in order to investigate the relationship between the disease, work performance and productive output.

2. THE EFFECTS OF S. MANSONI ON WORK PERFORMANCE IN CANE CUTTERS

An early attempt to measure work capacity directly in a working population with S. mansoni infection was made on a group of Sudanese cane cutters living in two villages in the Guneid sugar plantation on the Blue Nile (Collins et al. 1976). The investigations involved three stages: (a) a preliminary clinical and parasitological screening of about 500 cane cutters in order to identify infected and non-infected groups; (b) laboratory-based studies of working capacity and assessment of physique; and (c) studies on productivity in the cane fields on the same group of men performing a self-paced task (cane cutting). About 200 cane cutters took part in the study and these were divided into three groups: non-infected (no schistosome eggs excreted in stools), infected (egg counts in any number) and those who excreted eggs and in addition had clinical signs of infection (hepatosplenomegaly). The assumption was made that, in the absence of malaria and other parasitic diseases, cane cutters with hepatosplenomegaly were more seriously ill with S. mansoni infection. This is not, however, an ideal method of classification, for although many of the migrant cane cutters in the sample did not originate from endemic malarious areas, a proportion of them may have had malarial

splenomegaly in the absence of parasitaemia. Those infected
with urinary schistosomiasis (S. haematobium) were excluded,
as were cane cutters found to have malaria or other parasitic
(e.g., hookworm) infections, or with haemoglobin levels less
than 10 g/dl. Each of the three groups were then further sub-
divided into younger (16–24 years) and older (24–45 years) age
groups.


2.1.  Physiological responses to exercise

   Investigations were carried out on six groups as defined
above, each group consisting of about 30 cane cutters. Prior
to the exercise test on a bicycle ergometer, each subject was
allowed a 15 min practice period. The exercise consisted of a
progressive test, the work rate being raised every 3 min by
300 kpm/min from zero load to the limit of the individual's
work capacity. Minute ventilation, oxygen intake and cardiac
and respiratory frequencies were measured during the final
minute of each workload. Direct measurements of maximal
aerobic power output ($\dot{V}O_2$ max) were not attempted, but an
estimate was made from the individual regression equations of
oxygen consumption and heart rate. The results summarized
in table 1 show no group differences in exercise responses and
they accord with previous findings for East African sugar-cane
cutters who were free of parasitic infection and not anaemic
(Davies 1973). The mean submaximal exercise oxygen uptake
at a work output of 900 kpm/min ($\dot{V}O_2$ 900) and the predicted
maximal oxygen uptake at a cardiac frequency of 195 beats/min
($\dot{V}O_2$ 195) in both absolute terms and relative to body weight,
lean body mass and leg volume, were similar in the six groups
and independent of age and degree of schistosomiasis infection.
In view of the relationships between age and cardiac frequency,
calculation of $\dot{V}O_2$ max from $\dot{V}O_2$ 195 may lead to underestimation in the younger and overestimation in older subjects. As
a whole, the laboratory data gave no support for the view that
S. mansoni infection impairs physiological responses to a standard test of physical work as measured in adult males exercising
on a stationary bicycle ergometer.


2.2.  Productivity studies

   Individual measurements of productive output for about 130
cane cutters drawn from the six groups were then made in
terms of the actual weight of cane cut per unit time (kg/min)
during two separate days. The first day of the study simulated

Table 1. Oxygen uptake at a work output of 900 kpm/min ($\dot{V}O_2$ 900) and at a maximum cardiac frequency of 195 beats/min ($\dot{V}O_2$max) in absolute and relative terms (body weight, lean body mass, leg volume) in non-infected (ni), infected (i) and infected with signs (i+s) cane cutters (mean±S.D.)

| Group | n | $\dot{V}O_2$ 900 (l/min) | $\dot{V}O_2$max Abs (l/min) | $\dot{V}O_2$max Wt [(ml/kg)/min] | $\dot{V}O_2$max LBM [(ml/kg)/min] | $\dot{V}O_2$max LV [(ml/l)/min] |
|---|---|---|---|---|---|---|
| 16-24 y | | | | | | |
| ni | 28 | 2.23±0.16 | 2.88±0.43 | 48.3±5.7 | 53.9±6.2 | 230±40 |
| i | 40 | 2.16±0.22 | 2.84±0.57 | 50.7±7.1 | 55.7±7.7 | 240±30 |
| i+s | 31 | 2.26±0.19 | 2.83±0.32 | 49.8±5.4 | 55.1±6.1 | 240±30 |
| 25-45 y | | | | | | |
| ni | 27 | 2.13±0.19 | 2.67±0.42 | 47.3±6.4 | 52.0±6.8 | 230±40 |
| i | 41 | 2.20±0.14 | 2.80±0.49 | 46.4±7.6 | 51.6±8.0 | 230±40 |
| i+s | 27 | 2.21±0.16 | 2.97±0.70 | 49.3±8.7 | 54.4±9.6 | 240±40 |

*Energy and Effort*

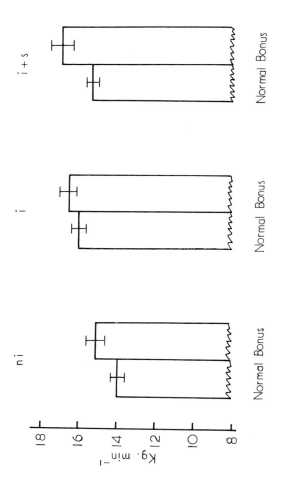

Figure 1. Mean productivity from S. mansoni infected [(i) and (i+s)] and non-infected (ni) cane cutters during normal working and bonus days. Productivity is expressed as (kg cane cut/man)/min (mean±S.D.)

a normal working day in order to indicate the cutters' habitual activity, and the second was designed to measure maximum productivity using a bonus system as inducement. The laboratory tests of exercise capacity (table 1) showed little difference between the younger and older subgroups, and the productivity results were therefore considered in relation to the degree of infection but not age (figure 1). It was found that the non-infected group were generally the lowest producers, the infected group the highest and the infected with clinical signs intermediate. The bonus system was effective in increasing output in all three groups and most especially in those who were infected and had clinical signs. The higher output from the infected (i) group as compared with the non-infected (ni) group on a normal working day basis was significant at the 5% level. Thus, it was found that under normal working conditions the infected group had a 10% higher productivity and the infected-with-signs group a 5% higher productivity than the non-infected. The most likely explanation for this paradoxical finding appears to be not in differences in aerobic capacity but in the different degrees of expertise of the groups. The average number of season's experience at cane cutting was found to be 1.4±1.5 for Group ni, 3.5±2.4 for Group i and 3.6±2.5 for Group i+s, and it appears that the development of habitual skill rather than intensity of infection with schistosomiasis is the more dominant factor. Experience at cane cutting was positively correlated at the 2% significance level with mean productivity.

## 2.3. Relative work rate

The energy cost (oxygen consumption) of cane cutting was recorded by Kofranyi-Michaelis portable respirometers. Assuming an RQ of 0.85, oxygen intake was converted into energy expenditure using the formula of Weir (1949) in which $CO_2$ concentration is not taken into account. The highest oxygen consumption during normal working was observed in the infected group (i) which accords with the higher productivity in this group noted above.

The oxygen uptake during work measured in 30 of the Guneid cane cutters ranged from 1.02 to 2.37 l/min, which was correlated (but not at the level of statistical significance) with laboratory measurements of $\dot{V}O_2$ max. However, in 42 villagers undertaking digging tasks with oxygen uptakes from 0.6 to 1.7 l/min, there was a significant correlation (p<0.001) between actual and potential energy expenditure (figure 2). This suggests that maximum aerobic capacity measurements in the laboratory provide an estimate of actual work performance in the

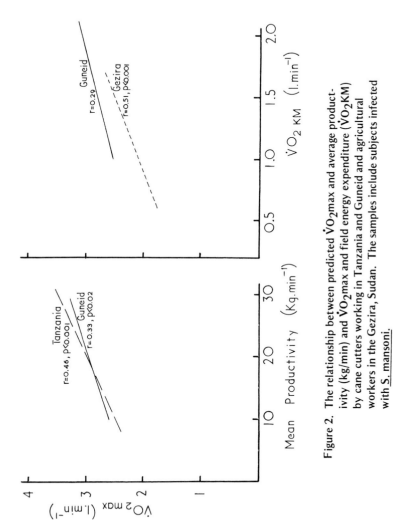

**Figure 2.** The relationship between predicted $\dot{V}O_2$max and average productivity (kg/min) and $\dot{V}O_2$max and field energy expenditure ($\dot{V}O_2$KM)
by cane cutters working in Tanzania and Guneid and agricultural
workers in the Gezira, Sudan. The samples include subjects infected
with <u>S. mansoni.</u>

Table 2. Predicted maximum oxygen uptake ($\dot{V}O_2$max), actual oxygen consumption ($\dot{V}O_2$ KM), heart rate (fH) and relative work rate (RWR) in non-infected and infected groups during cane cutting ($\pm$S.D.).

| | n | fH*<br>(beats/min) | $\dot{V}O_2$max<br>(l/min) | $\dot{V}O_2$ KM<br>(l/min) | RWR<br>(%) |
|---|---|---|---|---|---|
| Non-infected | 18 | $132\pm16$ | $2.76\pm0.50$ | $1.66\pm0.28$ | 60 |
| Infected | 13 | $139\pm15$ | $3.09\pm0.40$ | $1.80\pm0.38$ | 58 |
| Infected + signs | 10 | $134\pm9$ | $2.90\pm0.51$ | $1.45\pm0.12$ | 50 |

* Derived from the average of three to eight heart rates observed at half-hourly intervals

field. It was also found that the mean productivity (mean of normal and bonus days productivity) was also significantly correlated (at the 2% level) with the individual's estimated maximum aerobic capacity as shown in figure 2 (Collins et al. 1976). A very similar relationship between productivity and predicted $\dot{V}O_2$ max was found in East African workers employed at the Kilombero sugar cane estate (Davies 1973, Davies et al. 1976) in Tanzania.

An index of an individual's actual as compared with potential work performance may be deduced from the percentage of $\dot{V}O_2$ max, i.e., relative work rate, where

$$\text{Relative work rate} = \%\ \dot{V}O_2\text{max} = \frac{\text{Actual oxygen uptake in field work}}{\text{Laboratory estimate of } \dot{V}O_2\text{max}} \times 100$$

It was shown that, on average, the cane cutters at Guneid worked at about 57% of their $\dot{V}O_2$ max (range 34% to 76%). If the relative work rates of the three groups (ni, i and i+s) are considered, the non-infected worked at a higher relative work rate than infected cutters, though they had a slightly lower individual potential work capacity (table 2). These differences were not statistically significant, but if this trend and the productivity figures given in figure 1 are confirmed in more extensive studies it would suggest that though schistosomiasis might affect relative work rate, productivity is more directly related to the individual's potential work capacity.

Maximal performance predicted from submaximal exercise responses on a bicycle ergometer may underestimate $\dot{V}O_2$ max (Davies et al. 1976) and may therefore overestimate the relative workload as calculated above. Furthermore, it is not possible to reproduce the work of cane cutting, which is mainly confined to the upper limbs, by exercise on a bicycle ergometer. The overestimation of the relative work rate of cane cutters by bicycle ergometer tests may, however, be partly compensated for by the fact that $\dot{V}O_2$ max of unrestricted work with the upper body is lower than that of upright bicycle work (Åstrand & Saltin 1961).


3.  THE EFFECTS OF S. MANSONI ON ENERGY EXPENDITURE
    IN VILLAGE COMMUNITIES

In spite of infection with S. mansoni, the cane cutters at Guneid were fit enough to perform hard physical work with high average oxygen consumptions. Some of the cane-cutters were migrants from non-endemic areas of Sudan and mostly suffered from the effects of acute infection. A better index of the human and economic impact of the disease would appear to be

Table 3. Submaximal and maximal aerobic responses to exercise (means±S.D.) of non-infected and infected villagers and canal cleaners in the Gezira. Oxygen intake ($VO_2$) at a work output of 900 kpm/min ($VO_2$ 900) cardiac frequency at a $VO_2$ of 1.5/min ($fH_{1.5}$) and maximum $VO_2$ at an fH of (210 - 0.65 x age) beats/min ($VO_2$ max) in absolute (l/min) and relative terms [(ml/kg body weight)/min, (ml/kg lean body mass)/min and (ml/l leg volume)/min].

| | Non-infected villagers | Infected villagers | Canal cleaners |
|---|---|---|---|
| n | 37 | 146 | 19 |
| Age (y) | 26.2±5.9 | 27.1±7.0 | 27.4±6.8 |
| Hb (g/dl) | 15.2±1.3 | 14.9±1.2 | 13.3±2.2 |
| Egg excretion (per g faeces) | 0 | 367±486 | 2054±1105 |
| $VO_2$ 900 (l/min) | 2.24±0.16 | 2.21±0.18 | 2.17±0.16 |
| $fH_{1.5}$ (beats/min) | 142±15 | 139±16 | 148±16 |
| $VO_2$ max Abs (l/min) | 2.71±0.71 | 2.63±0.56 | 2.39±0.46 |
| $VO_2$ max Wt [(ml/kg)/min] | 46.9±12.0 | 47.4±10.7 | 41.9±5.7 |
| $VO_2$ max LBM [(ml/kg)/min] | 53.9±14.0 | 54.2±12.5 | 46.3±6.8 |
| $VO_2$ max LV [(ml/l)/min] | 247±56 | 251±55 | 207±29 |

obtained from static and less physically well-trained populations, and therefore the next studies were made on settled village communities inhabiting the Gezira irrigated area between the two Niles (Awad El Karim et al. 1980).
The study was carried out in an area with high S. mansoni incidence where the population had received no mass anti-schistosomal treatment and where the irrigation canals were untreated with molluscicide. At the time of the investigation, the prevalence of malaria in the Gezira was low (0.1 – 1%) and of a stable type. Nevertheless, an anti-malarial prophylaxis programme was arranged for the villagers to ensure that none of the subjects tested was suffering from malaria. The methods were essentially the same as those used at Guneid, but this time more careful consideration was given to the role of quantitative egg excretion in determining the level of work capacity. One of the reasons for this was the recent demonstration that, in children at least, a relationship can be shown between objective morbidity (hepatosplenomegaly) and intensity of infection (quantitative egg excretion) (Cook et al. 1974). Only the economically active males (18–45 year old) were included in the study, and of these about 20% were not infected. Submaximal exercise tests to determine $\dot{V}O_2$ max (predicted) and in some cases direct measurements of $\dot{V}O_2$ max were performed by bringing the villagers to a central laboratory. The results (table 3) show that there were no significant differences in submaximal exercise responses or in maximum aerobic capacity measured both in absolute terms and in relation to body weight, lean body mass and leg volume. This virtually confirmed the findings in the laboratory tests on the Guneid cane cutters.

4.    WORK PERFORMANCE WITH HIGH LEVELS OF S. MANSONI INFECTION

A third group of people who had lived on average for 14 years in the Gezira villages were also selected for work-capacity tests. These were canal cleaners who were all heavily infected by S. mansoni, and who by the nature of their occupation were forced to risk reinfection daily. The diet of the canal cleaners was similar to that of the other villagers, and though as a group their income was lower, the additional support usually given by other members of their families suggested that there were no significant differences in living standards compared to other villagers. For statistical analysis of the data shown in table 3, the Newman-Keuls studentized range was used for post-hoc tests on differences between the group means. The

$\dot{V}O_2$900 of the canal cleaners was lower than that for the villagers (infected and non-infected) but the difference is not statistically significant. In absolute terms, the mean predicted maximum aerobic power output was lower in the canal cleaners as was also the $\dot{V}O_2$max per kg body weight, but again these differences were not significant. However, the differences between villagers and canal cleaners in terms of $\dot{V}O_2$max per kg lean body mass and in relation to leg muscle (plus bone) volume were found to be significant at the 5% and 0.2% levels respectively.

It seems therefore that a high level of intensity of infection with schistosomiasis might after all be found to influence physical performance. Work capacity does not appear to be impaired in lightly infected villagers with egg counts less than 1000 eggs/g, which is in accord with previous findings on the cane cutters at Guneid. However, impairment in physical work capacity becomes significant when the pattern of infection changes from light to very heavy (>2000 eggs/g) where, as in the example of the canal cleaners, the level of impairment of $\dot{V}O_2$max (relative to leg muscle volume) may be as much as 18%. It may be argued that the factor which affects this result is the special working conditions of the canal cleaners who were often immersed to the waist in water. It is conceivable that this might cause detraining of the leg muscles for work tests on a bicycle ergometer. However, the physical characteristics, including height, lean body mass and leg muscle volume, were actually slightly greater in the canal cleaners than in the other villagers.

Table 3 also illustrates that there was a significant reduction in haemoglobin level (P<0.001) in canal cleaners, and this might afford an explanation for the effect of schisotosomiasis on work capacity. Anaemia in schistosomiasis is reported to be of the haemolytic type initiated by an immune process (Mahmoud & Woodruff 1972, Woodruff 1973), though a hypochromic anaemia due to chronic blood loss might also contribute (Ongom & Bradley 1972). In retrospect, it was the results on the Guneid cane cutters which helped to confirm the conclusion that intense infections impaired work performance. Quantitative egg excretion rates had been measured in some of the Guneid cane cutters, and it was found that two had very high egg loads, 3500 and 3800 eggs/g respectively. These two also had very low values for predicted $\dot{V}O_2$max, 1.84 and 1.86 l/min, which were strikingly lower than the average for the infected cane cutters and also low haemoglobin levels, 10.9 and 12.9 g/dl.

## 5. THE EFFECT OF TREATMENT ON POTENTIAL WORK PERFORMANCE

Cook et al. (1977) have evaluated different methods for controlling S. mansoni transmission, and have suggested that in some ecological situations chemotherapy can produce a more striking reduction in the prevalence and incidence of schistosomiasis than either mollusciciding or water-supply control. Apart from its conventional therapeutic use, schistosomicidal treatment might also be expected to have a significant effect on the work output of infected individuals, if indeed S. mansoni infection does have a deleterious effect on performance. An investigation was therefore made on a group of infected Gezira villagers in order to measure the effect of hycanthone treatment on the physiological response to graded physical exercise, while a control group, infected but untreated, was tested under similar conditions (Awad El Karim et al. 1981).

After initial bicycle ergometer progressive exercise tests, 22 infected males (18–45 years of age) were given 3 mg hycanthone per kg body weight as a single intramuscular injection and admitted for 2–3 days observation in hospital. Following discharge from hospital, the subjects were instructed to avoid contact with any possible source of schistosomiasis infection and, in addition, were given chloroquine anti-malarial prophylaxis. During the following year, the subjects were monitored periodically to ensure that they had remained free of schistosomiasis and were then retested on the bicycle ergometer. Another 19 infected males from the same villages were tested in the same way as a control group, receiving anti-malarial prophylaxis but no schistosomicidal treatment.

Submaximal exercise tests in the two groups showed that there was little change in $\dot{V}O_2900$ after treatment. There were, however, post-treatment increases in maximal aerobic power in the treated group measured both in absolute terms and in relation to body weight, lean body mass and leg volume (table 4). These changes did not occur in the untreated group. Subjectively, about half of the treated subjects claimed that they suffered far less fatigue in everyday work than they did before treatment. There was no such improvement reported by the untreated subjects. Schistosomicidal treatment was also associated with a greater increase in haemoglobin concentration and in forced vital capacity (FVC) in the treated group.

Similar longitudinal studies are now in progress in Gezira villages in order to assess the effect of schistosomicidal treatment on performance and habitual energy expenditure during standardized agricultural work tasks.

Table 4. Effects of schistosomicidal treatment on physiological responses to exercise (mean values ±S.D.). $VO_2$max was predicted from a maximum cardiac frequency of 210 (beats/min)−0.65 x age and expressed in absolute values and in relation to body weight (kg), lean body mass (kg) and leg volume (l). Forced expiratory volume in 1 sec ($FEV_1$) and forced vital capacity (FVC) were measured by vitalograph.

| | Untreated (n=19) | | Treated (n=22) | |
|---|---|---|---|---|
| | 1st Test | 2nd Test | 1st Test | 2nd Test |
| Age (y) | 29.5±8.1 | 29.9±8.1 | 25.4±5.3 | 26.4±5.3 |
| Lean body mass (kg) | 50.1±6.6 | 50.9±6.0 | 47.7±5.0 | 50.1±6.9 |
| Haemoglobin (g/dl) | 14.8±1.3 | 15.3±1.1 | 14.1±0.9 | 15.2±0.9 [1,2] |
| Egg excretion (eggs/g) | 257±268 | 445±394 [1] | 526±428 | 0 [1,2] |
| $FEV_1$ (l/sec) | 3.3±0.6 | 3.4±0.6 | 3.7±0.4 | 3.7±0.4 |
| FVC (l) | 4.5±0.8 | 4.3±0.7 | 4.2±0.5 | 4.5±0.5 [1,2] |
| $VO_2$ 900 (l/min) | 2.2±0.2 | 2.1±0.1 | 2.2±0.3 | 2.1±0.2 |
| $VO_2$ max Abs (l/min) | 2.7±0.7 | 2.6±0.6 | 2.4±0.5 | 2.8±0.7 [1,2] |
| $VO_2$ max Wt [(ml/kg)/min] | 48.9±12.6 | 45.5±8.1 | 46.6±12.0 | 49.8±10.0 |
| $VO_2$ max LBM [(ml/kg)/min] | 56.5±12.7 | 51.4±8.4 | 51.4±12.9 | 55.8±11.2 |
| $VO_2$ max LV [(ml/l)/min] | 269.4±45.6 | 252.1±75.8 [1] | 225.1±49.9 | 245.8±55.4 [1,2] |

[1] Significant changes differing from zero   [2] Significant changes associated with treatment

6. DISCUSSION

The investigations in the Sudan on energy expenditure and schistosomiasis represent an attempt to quantify the relationship between disease intensity and work productivity by combined laboratory and field studies. It cannot be claimed at this stage that these methods can be directly applied to assess the economic damage caused to communities by schistosomiasis. Nevertheless, it has been argued (Prescott 1979) that attempts to estimate the effects of schistosomiasis by computing annual economic losses will have to rely eventually on the empirical basis established by investigations on small population groups. It is perhaps not surprising that there is widespread acceptance that "schistosomiasis not only decidedly and often profoundly affects the physical capacity of the majority of infected persons, but also lowers their resistance to other infections, hence producing a marked reduction in their productive capacity", as expressed in a WHO monograph. Most attempts to estimate the value of increasing output by eliminating schistosomiasis imply huge economic benefits (e.g., Khalil 1949, Farooq 1967, Wright 1972) and in mainland China, where the pathogenic S. japonicum is endemic, it is claimed that the disease can cause an average loss of 40% of an adult's capacity to work (Cheng 1971). These claims, however, do not have universal support. For example, in an attempt to relate changes in the prevalence of schistosomiasis to changes of productivity in terms of rice yields in China, Andreano (1976) could identify no clear relationship. Controlled physiological studies which have been conducted on children (Davies 1972, Walker et al. 1972) and the present investigations on agricultural workers and cane cutters in Sudan give little support to the view that potential work capacity is impaired, at least with moderate schistosome infections. Our studies suggest that it is only the minority of the active working population with very high schistosome egg loads who demonstrate any significant impairment. For the most part, infected workers appear able to work as well as the non-infected, given the same degree of skill, and this suggests that there may be a threshold level of infection below which there is little or no apparent detrimental effect.

Whatever degree of confidence can be given to economic estimates of the consequences of an endemic disease such as schistosomiasis, it is important to consider how the disease affects active work performance. The ability to work hard may be unimportant in work which does not demand physical effort. On the other hand, psychological factors and skill contribute to both active and sedentary work performance, and the possible effects of schistosomiasis on mental effort and motivation requires

further investigation. Analysis of the earnings of sugar estate
workers in Tanzania (Fenwick & Figenschou 1972) showed that
non-infected workers earned at least 11% more in bonuses than
those with S. mansoni infection. The significant decrease in
productivity of those infected, which was of the order of 5%,
was found to be associated with a higher incidence of absenteeism. This, therefore, is probably the primary cause of low
productivity – loss of output because the worker is too ill to
attend work. If work is continuously available, and in many
developing countries this is by no means normal, the worker
suffering from schistosomiasis may compensate for reduced
vigour by working longer hours or receiving extra help from
others. A worker who is chronically ill and whose vigour has
been impaired for some time may suffer detraining effects and
is likely to adopt different working habits, habits that offer
the least stress. The presence of changed work patterns such
as this helps to confound calculations based on annual productivity figures.

To try to identify the root cause of the effects of schistosomiasis on physiological work capacity, one has initially to consider the impact of a disease, contracted often in early childhood, on the growth and development of the individual. The
total power developed by the human body depends on its size,
and a disease that may stunt growth will certainly restrict the
full development of working capacity. In this respect, there
may be similarities to the long-term effects of malnutrition and
low haemoglobin levels. The deleterious effect of even mild
anaemia on physical work capacity has been demonstrated both
clinically and experimentally (Viteri & Torun 1974). There are
compensatory mechanisms, such as an increased circulation of
blood through energy-demanding tissues and more efficient
removal of available oxygen from haemoglobin, which sometimes
enable anaemic individuals to engage in occupations demanding
heavy energy expenditures. Behavioural patterns also play an
important role. Nevertheless, anaemia becomes functionally
limiting and relatively small decreases in haemoglobin level can
diminish performance at near-maximum levels of exercise. The
lowered haemoglobin concentration in the cane cutters and canal
cleaners with high schistosome egg-excretion rates may at least
partly explain the observed decreases in the laboratory-
measured potential work capacity. Low haemoglobin values in
subjects with high schistosome egg loads has also been reported
by Nelson (1958) and the mechanism of the haematological damage in schistosomiasis has been investigated by Mahmoud &
Woodruff (1972). It must be stressed, however, that low blood-
haemoglobin concentration may not be the only or even principal

functional component of the oxygen transport system affected by the disease. It appears likely that the lowered haemoglobin levels may be one important underlying cause of the fatigue, debility, disinclination to work and poor physical performance in severe cases of schistosomiasis, and it may also be a common factor in the tropical anaemias and other widespread tropical diseases such as malaria which are considered to have a major influence on work, productivity and economic development (Pan American Health Organisation 1975).


7.  ACKNOWLEDGEMENTS

    I would like to acknowledge the support given by Professor J.S. Weiner throughout these studies, and collaboration in the Sudan by Dr. M.A. Awad El Karim and staff of the Departments of Physiology, Medicine and Community Medicine in the Faculty of Medicine, Khartoum University. The work has been funded by the Medical Research Council and the Edna McConnell Clark Foundation, New York.


REFERENCES

Abel-Smith, B., 1972, Health priorities in developing countries: the economist's contribution. International Journal of Health Services, 2, 5–12.
Andreano, R.L., 1976, The recent history of parasitic disease in China: the case of schistosomiasis, some public health and economic aspects. International Journal of Health Services, 6, 53–68.
Åstrand, P.-O. & Ryhming, I., 1954, A nomogram for calculation of aerobic capacity (physical fitness) from pulse rate during submaximal work. Journal of Applied Physiology, 7, 218–221.
Åstrand, P.-O. & Saltin, B., 1961, Oxygen uptake during the first minutes of heavy muscular exercise. Journal of Applied Physiology, 16, 971–976.
Awad El Karim, M.A., Collins, K.J., Brotherhood, J.R., Dore, C., Weiner, J.S., Sukkar, M.Y., Omer, A.H.S. & Amin, M.A., 1980, Quantitative egg excretion and work capacity in a Gezira population infected with Schistosoma mansoni. American Journal of Tropical Medicine and Hygiene, 29, 54–61.
Awad El Karim, M.A., Collins, K.J., Sukkar, M.Y., Omer, A.H.S., Amin, M.A. & Dore, C., 1981, An assessment of anti-schistosome treatment on physical work capacity. Journal of Tropical Medicine and Hygiene, 84, 65–70.

Cheng, T.-H., 1971, Schistosomiasis in mainland China. A review of research and control programs since 1949. American Journal of Tropical Medicine and Hygiene, 20, 26-53.
Collins, K.J., Brotherhood, J.R., Davies, C.T.M., Dore, C., Hackett, A.J., Imms, F.J., Musgrove, J., Weiner, J.S., Amin, M.A., Awad El Karim, M.A., Ismail, H.M., Omer, A.H.S. & Sukkar, M.Y., 1976, Physiological performance and work capacity of Sudanese cane cutters with Schistosoma mansoni infection. American Journal of Tropical Medicine and Hygiene, 25, 410-421.
Cook, J.A., Baker, S.T., Warren, K.S. & Jordan, P., 1974, A controlled study of morbidity of Schistosoma mansoni in St. Lucian children, based on quantitative egg excretion. American Journal of Tropical Medicine and Hygiene, 23, 625-633.
Cook, J.A., Jordan, P. & Bartholomew, R.K., 1977, Control of Schistosoma mansoni transmission by chemotherapy in St. Lucia. American Journal of Tropical Medicine and Hygiene, 26, 887-893.
Davies, C.T.M., 1972, The effects of schistosomiasis, anaemia and malnutrition on the responses to exercise in African children. Journal of Physiology, 230 27P.
Davies, C.T.M., 1973, Relationship of maximum aerobic power output to productivity and absenteeism of East African sugar cane workers. British Journal of Industrial Medicine, 20, 146-154.
Davies, C.T.M., Brotherhood, J.R., Collins, K.J., Dore, C., Imms, F., Musgrove, J., Weiner, J.S., Amin, M.A., Ismail, H.M., Awad El Karim, M.A., Omer, A.H.S. & Sukkar, M.Y., 1976, Energy expenditure and physiological performance of Sudanese cane cutters. British Journal of Industrial Medicine, 33, 181-186.
El Batawi, M.A., 1975, Health of working populations in industrializing societies. In Health and Industrial Growth, CIBA Foundation Symposium No. 32 (Amsterdam: Associated Scientific Publishers), pp. 141-156.
Farooq, M., 1967, Progress in bilharziasis control: Egypt. Chronicle of the World Health Organization, 21, 175-184.
Fenwick, A. & Figenschou, B.M., 1972, The effect of Schistosoma mansoni on the productivity of cane cutters on a sugar estate in Tanzania. Bulletin of the World Health Organization, 47, 567-572.
Harrison, M.H., Brown, G.A. & Cochrane, L.A., 1980, Maximal oxygen uptake: its measurement, application, and limitations. Aviation, Space and Environmental Medicine, 51, 1123-1127.
Khalil, M., 1949, The national campaign for the treatment and control of schistosomiasis from the scientific and economic

aspects. Journal of the Royal Egyptian Medical Association, 32, 817–856.

Mahmoud, A.A.F. & Woodruff, A.W., 1972, Mechanisms involved in the anaemia of schistosomiasis. Transactions of the Royal Society of Tropical Medicine and Hygiene, 66, 75–84.

Margaria, R., Aghemo, P. & Rovelli, E., 1965, Indirect determination of maximal oxygen consumption in man. Journal of Applied Physiology, 20, 1070–1073.

Nelson, G.S., 1958, Schistosoma mansoni infection in the West Nile district of Uganda, Part IV. Anaemia and S. mansoni infection. East African Medical Journal, 35, 581–586.

Ongom, V.L. & Bradley, D.J., 1972, The epidemiology and consequences of Schistosoma mansoni infection in West Nile, Uganda. Transactions of the Royal Society of Tropical Medicine and Hygiene, 66, 835–851.

Pan–American Health Organization, 1975, The impact of malaria on economic development: a case study. Regional Office of the World Health Organization, Scientific Publication No. 297.

Prescott, N.M., 1979, Schistosomiasis and development. World Development, 7, 1–14.

Shephard, R.J., 1980, Population aspects of human working capacity. Annals of Human Biology, 7, 1–28.

Viteri, F.E. & Torun, B., 1974, Anaemia and physical working capacity. Clinics in Haematology, 3, 609–626.

Walker, A.R.P., Walker, B.F., Richardson, B.D. & Smit, P.J., 1972, Running performance in South African Bantu children with schistosomiasis. Tropical Geographical Medicine, 24, 347–352.

Weir, J.B. de V., 1949, New methods for calculating metabolic rate with special reference to protein metabolism. Journal of Physiology, 109, 1–9.

Weisbrod, B.A., 1961, Economics of Public Health. Measuring the economic impact of diseases (Philadelpha: University of Philadelphia Press).

Woodruff, A.W., 1973, Mechanisms involved in anaemia associated with infection and splenomegaly in the tropics. Transactions of the Royal Society of Tropical Medicine and Hygiene, 67, 313–328.

World Health Organization, 1973, Schistosomiasis control. Report of a WHO expert committee. Technical Report Series No. 515.

Wright, W.H., 1972, A consideration of the economic impact of schistosomiasis. Bulletin of the World Health Organization, 47, 559–566.

# THE COMPARISON OF ENERGY SUPPLY AND ENERGY NEED:
## A CRITIQUE OF ENERGY REQUIREMENTS

J.P.W. Rivers and P.R. Payne

Department of Human Nutrition,
London School of Hygiene and Tropical Medicine,
Keppel Street,
London WC1E 7HT

## 1. INTRODUCTION

There are many published comparisons in the literature between food supplies and food needs, and the production of such estimates is an activity that has grown progressively in importance in the last 50 years. These estimates have been based almost completely on the comparison of nutrient requirements with statistics on agricultural production rather than with real food-intake data.

National statistics on agricultural production are obviously of variable quality, but discussion of their strength or demerits is outside our competence, and, as long as they are used simply to provide an index of the viability of one section of the economy, they are outside the interests of this symposium.

Unfortunately, such statistics have been coupled with demographic and physiological data in attempts to provide information on the adequacy of food supplies for populations, and most unfortunately of all, some of these analyses purport to give information on the extent of nutritional deficiencies in the world.

When The Economist magazine (1952) savagely characterized the FAO as 'a permanent organization devoted to proving that there is not enough food in the world' it was referring to its leading role in this process. Consistently, FAO and its followers have coupled food-production data with more or less fictitious estimates of storage losses and preparation losses, and divided by population estimates (also of doubtful quality) to produce estimates of per-capita food supplies.

These aggregated data, which take no allowance of the fact that food supplies vary between the members of a society, have then been compared with estimates of average physiological need to confirm a number of a priori conceptions on malnutrition which existed for reasons that had little to do with the evidence. During the period 1950-1970 such statistics were used to show that the world was short of protein, whereas from about 1965-1975 they were interpreted as showing that the world was short of food energy. They

continue to be used to persuade us that the problem of under-
development should be described in terms of deficits in food
supply, obviating the need to consider it in terms such as poverty
and exploitation.  In this way, our attention is directed to a
comfortable and reassuring solution, the paternalism of aid,
rather than the political turmoil of reassessing the nature of our
own economic system and its role in restricting the economic
development of other nations.

In this paper we shall focus on the role of so-called nutrient
requirements in this process and argue that they have played a
major role in generating misconceptions about the causes and
extent of hunger.  We suggest that the continuing process of
sophisticating and revising them by a succession of expert committees only serves to sustain such myths.

We shall argue that the agencies and institutions which commission and fund these committees frequently manipulate the
expert members so as to achieve answers which they require for
their own political purposes:  the 'experts', for their part,
generally fail to perceive the extent of that manipulation, or the
extent to which their scientific judgement is founded upon their
own political and social viewpoints.

One particular outcome of these long-sustained incestuous
relationships between administrators and their professional advi-
sors has been their adherence to a particular 'model' or conceptual
framework.  This model is rarely explicitly described, but underlies all requirement estimates to date.

This model assumes that once all obvious sources of difference
between individuals - age, sex, body size and physical activity -
have been allowed for, remaining differences between people are
constitutional or genetically determined.  In addition, these
differences in needs are fixed in value so that the requirement
for a group of people is completely specified by the mean and
standard deviation of a 'normal' distribution.  Each individual
has his own particular level of need within that distribution, and
if his intake fails to match, he will be regarded as 'abnormal' or
'malnourished'.

This particular model of what requirements are, and what happens when they are not met, is not clearly admitted as an assumption underlying most expert committee reports.  Neither has it,
to our knowledge, been subjected to experimental tests.

In fact, of course, other models could be proposed which are
both plausible and testable.  In particular, an adaptive model
seems inherently more likely:  individuals may adapt safely and
reversibly to a range of intakes with perfectly acceptable changes
in functional performance.  If pushed beyond that range, however, adaptation may no longer be possible without performance
being reduced.  In this model, only the upper and lower limits

to adaptability may be genetically determined.

We believe that, if the adaptive model were used as a basis for estimates of requirements, these could be used, for example, to define a lower level of subsistence below which starvation was inevitable. However, such values would be of limited use, except as an extreme index of poverty. We would not be able to use them to prescribe adequate diets nor to assess the extent of shortfalls in food supply. If our concern is to achieve equity in economic development, then the conventional physiologists' 'model' of nutrient requirements, and the present machinery for sanctifying the values based on that model, is at best irrelevant and at worst a force for reaction.

## 2. NUTRIENT REQUIREMENTS

If the science of nutrition has any central concept, it is surely the notion of nutrient requirements. The initial impetus for the subject came from 19th century attempts to define food needs; and personal pronouncements on the quantitative nature of protein and/or energy requirements can be associated with almost all the founding fathers of nutrition. This is to be expected. A science concerned with the physiological aspects of diet must make some statements about the nature of food needs to achieve credibility.

What is less predictable is the fact that, nearly 200 years after it began, the process of revising such requirements continues unabated. Nutritionists no longer regard themselves simply as a particular type of physiologist, and nutritional research ranges from the molecular to the political. Nevertheless, the problem of requirements remains apparently unresolved, in that it continues to occupy a niche in the writings of eminent nutritionists, and the rate of production of new estimates of nutrient requirements is now higher than ever before.

What has changed since the 19th century is that the estimation of nutrient requirements is no longer a matter of pronouncement made by eminent individual scientists. In keeping with the general trend towards 'Big Science' (De Solla Price 1963), estimates of nutrient requirements are now made by committees of experts convened by national academies, governments or international agencies, and published by these official bodies.

Although the resultant reports are useful review articles and text books, their social function is much wider than this. They are consulted by an audience that is much broader than nutritional physiologists, as indeed it is intended they should be. The problem is that amongst this group of users, there are many who

tend to apply the figures as if they possessed the precision and
the authority of a table of physical constants.

The reputation and influence that such reports on requirements
have must not be underestimated.   Nor must the fact that this
reputation does not result simply from the acceptance of their
scientific merit;  it is imposed mainly by the prestige associated
with their official status.

The authors of official reports do not seriously attempt to
disown this authority.   While, admittedly, the current FAO/WHO
report (FAO/WHO 1973), for example, does equivocate about the
basis of the figures advanced and their use and their accuracy,
such disclaimers are brief, often ambiguous and scattered
obscurely through the text.

But the title "Energy and Protein Requirements" carries no such
overtone of academic prevarication and the tables of values it
contains, which are its inevitable focus, generate an atmosphere
of certainty.   Words like 'guess', 'ignorance' or 'estimate' are
noticeably rare and the overall impression of any hesitation in
the report is that, like a tradesman's disclaimer, it serves to
protect the producer's reputation, not the consumer's interest;
caveat emptor.

This combination of unchallengeable authority and uncritical
acceptance makes it necessary to consider carefully the nature
of the 'nutrient requirements' that are contained in official
reports and to distinguish carefully between the general concept
of nutrient requirements and the particular values of the estimates
for these which expert committees publish.

Obviously, all published values for scientific constants are in
fact only estimates of those constants, and most scientists are
aware that the published value for the coefficient of expansion
of brass, for example, is merely the currently favoured value,
to be displaced when technical improvements result in a more
precise estimate, or if theoretical improvements lead to the
revision, or even the abandonment, of the concept itself.
Generally, this restriction is implicit in the scientific method and
can be ignored.

But it is particularly important to differentiate specific estimates
from general concepts, and to define the nature of the concept
when nutrient requirements are considered, for two reasons.
Firstly, there is the problem to which we have referred, of the
plethora of official bodies that issue such requirements. Secondly,
there is the problem that the concept of a requirement is itself
intrinsically less tangible than the apparently straightforward
name suggests, and particular estimates may therefore have very
restricted validity as well as limited accuracy.

## 3. THE EXPERT COMMITTEE

Tables 1 and 2 give some of the results of the considerations of Expert Committees. Table 1 contains examples of the official national estimates for energy requirements, or recommended intakes of energy, current in 1975—76. They are taken from two recent attempts to collate these estimates: a round table held at a recent European Nutrition Congress (Wretlind 1977), and some comparisons made by a committee of the International Union of Nutritional Sciences (IUNS 1975).

Table 2 gives sequential estimates produced by some national expert committees and also by the United Nations/League of Nations agencies.

We suspect that a critical and comprehensive review of these would be a valuable, but disturbing, study of the sociology of science. None has yet been made. If it were, it is probable that it would conclude, as we do, that the large number of estimates made and the variability between them are due mainly to the political exigencies at different times and places rather than to differences of technique or of theory.

Large numbers of estimates are not invariably to be deplored in science. Obviously, successive estimates are made of any physical constant as accuracy of estimation improves. But in this case, there would be increases in precision, and sequential changes in the estimates will tend to converge within narrower and narrower limits. Expert committees on nutrient requirements do differ in the apparent precision of their estimates, some, if their tables are to be taken literally, giving values accurate to 0.01%, others to 1%. However, there is no real temporal trend in this, and no evidence to justify the apparent standards of precision adopted by any committee. Its basis seems to be inscrutable.

In some countries, time-trends can be discerned, though they hardly indicate convergence. The official USA figures, for example, as Poleman (1977) has pointed out and as table 2 makes clear, have shown a 16% fall in the energy requirements of adult males since 1958. This is not associated with improvements in the process of estimation. It is difficult to believe that it can be accounted for by factors such as changes in activity. If it represents physiological reality, then it must cause considerable concern, since if the trend continues unabated, USA citizens will cease to have an energy requirement by the year 2055, the laws of thermodynamics having presumably been repealed unilaterally by the USA.

Clear evidence of the lack of temporal convergence, and perhaps the most unequivocal indictment of modern estimates, comes from comparing the pronouncements of modern expert committees with the views of 19th and early 20th century experts. Their estimates, a

TABLE 1.    Estimates of energy requirement, or recommended
            allowances for energy (a) : official values in
            different nations 1975—1976.   Values are applicable
            to adult males 25 years old, moderately active and/or
            sedentary.   Where values are adjusted for body
            weight, a value of 65 kg has been used except where
            noted.

| Country | | Energy requirement (a) (kcal/day) | |
|---|---|---|---|
| | | Sedentary | Moderately active |
| Australia | | 2800 | – |
| Canada | | 2650 | – |
| Colombia | (f) | 2850 | – |
| Czechoslovakia | (f) | 2700 | – |
| Denmark | | (b) | (b) |
| Eire | | (c) | (c) |
| Finland | (f) | 2400 | 3000 |
| France | (f) | – | 3000 |
| German Democratic Republic | (f) | 2700 | 3000 |
| German Federal Republic | (f) | 2550 | – |
| Hungary | (f) | 2400 | 2700 |
| India | (f) | 2400 | – |
| Indonesia | (f) | 2600 | – |
| Italy | | 2700 | 3000 |
| Japan | (f) | 2500 | – |
| Malaysia | (f) | 2500 | – |
| Netherlands | (e) | 2600 | 2900 |
| Norway | (e) | 2800 | 3000 |
| Philippines | (f) | 2500 | – |
| Poland | (f) | 2600 | 3200 |
| Romania | (f) | 2500 | 3500 |
| Spain | | 2700 | 3000 |

| Country | | Energy requirement [a] (kcal/day) | |
|---|---|---|---|
| | | Sedentary | Moderately active |
| Sweden | (e) | – | 2800 |
| Thailand | (f) | 2550 | – |
| Turkey | (f) | 3000 | – |
| UK | | 2700 | 3000 |
| USA | (e) | 2800 | – |
| USSR | (e) | – | 3000 |
| Yugoslavia | | (d) | (d) |

Sources: IUNS (1975) and Wretlind (1977)

Notes:

(a) Since both overconsumption and underconsumption of energy is assumed to be harmful, expert committees currently equate the recommended allowance for energy with the mean requirement. In this paper, therefore, the values are regarded as interchangeable.
(b) Official policy is to adopt values for the USA.
(c) Official policy is to adopt values for the UK.
(d) Official policy is to adopt FAO/WHO values.
(e) Body weight of 70 kg assumed.
(f) No body-weight corrections made or available.

TABLE 2. Sequential estimates of energy requirements by
official bodies (a). Values cited are those applicable
for a moderately active 25 year old male: assumed
body weights are given

---

USA
Values are those produced by committees of the National Academy
of Sciences and published by the National Academy of Sciences
in the year shown. Assumed body weight is 70 kg.

| | |
|---|---|
| 1943 | 3000 |
| 1945 | 3000 |
| 1948 | 3000 |
| 1958 | 3200 |
| 1964 | 2900 |
| 1968 | 2800 |
| 1974 | 2700 |

India
Values are those promulgated by the Indian Council of Medical
Research in the year shown. Assumed body weight is 55 kg.

| | |
|---|---|
| 1944 | 2800 |
| 1960 | 3000 |
| 1968 | 2800 |
| 1971 | 2800 |

UK
Official bodies vary and are listed in references.
Body weight where stated is assumed to be 65 kg.

| | | | |
|---|---|---|---|
| 1933 | 3400 | BMA | 1933 |
| 1950 | 3200 | BMA | 1950 |
| 1969 | 3000 | DHSS | 1969 |
| 1979 | 2900 | DHSS | 1979 |

United Nations/League of Nations
Sources are as given in references.
Where body weight is given values refer to 65 kg.

| | | |
|---|---|---|
| 1918 | 3000 | Inter-Allied Scientific Commission 1918 cited in FAO/WHO 1973 |
| 1935 | 3000 | League of Nations Health Organization 1935 |
| 1950 | 3200 | FAO  1950 |
| 1957 | 3200 | FAO  1957 |
| 1973 | 3000 | FAO/WHO 1973 |

---

(a) See note (a), table 1.

TABLE 3. Some estimates of energy requirements made by physiologists before 1914. Values are for adult males in Europe or the USA. Average body weights are assumed, where stated values range from 60—70 kg.

| Source and date | Energy requirement [b] (kcal/day) | | |
|---|---|---|---|
| | Light activity | Moderate activity | Hard work |
| Moleschott 1859 | | 3160 | |
| Lyon Playfair 1865[a] | 3029 | 3146 | 4060 |
| Ranke 1876 | | 3195 | 3574 |
| Studemund 1878 | | 3229 | |
| Voit 1881 | | 3055 | 3370 |
| Hultgren & Landegren 1891 | | 3436 | 4726 |
| Atwater 1895 | 3000 | 3500 | |
| Rubner 1902[a] | 2631 | 3121 | 3644 |
| Lichtenfelt 1903 | | 2700 | 3088 |
| Schmidt 1901 | | 3235 | |
| Gautier 1907 | | 2830 | 4247 |
| Chittenden 1907 | | 2800 | |
| Lusk 1909 | | 3000 | |

(a) Values cited from Hutchinson 1906 (Chapter 2).

(b) Some of the variation in energy requirements at different work levels will be due to idiosyncrasies in definition of work level by different authors.

representative selection of which are given in table 3, look suspiciously similar to those of the 20th century committees. In fact, there is no significant difference between the mean values in table 3 and those in tables 1 or 2. Moreover, there is a depressing similarity in methodology, Voit's (1881) 'standard', for example, being based on observed food intake, as is the comparable one of FAO/WHO (1973).

It is clearly important to understand, therefore, what all these expert committees are really doing when they produce their reports; since the reports themselves fictionalize the process, citing the evidence they use to deduce their view, but not explaining why so much has been ignored.

This process is common to all scientific publications, and if scientists had to cite all the evidence which related to a topic, scientific papers would become books and conclusions would be impossible. The process of rejecting the work you don't happen to like is intrinsic to the structure of science and is absolutely necessary (Kuhn 1970). But it means that scientific theories are created by scientists out of a partial view of 'reality'. And it means that the views of even the most exalted expert committees represent only the sum of the pre-existing prejudices of its members.

The clearest illustration of this comes from the field of protein requirements. During the periods when expert committees recommended high protein diets and thus manufactured a world protein gap, Chittenden's (1905) experimental demonstration that human requirements were, in fact, much lower was ignored. Indeed, it is impossible to deduce from the reports that Chittenden ever even lived, let alone that he published several books and many papers on the subject of the nutrition of man (see, for example, Chittenden 1907).

With the reduction of estimates of protein requirements by the FAO/WHO Committee in 1971 (FAO/WHO 1973) Chittenden's work was disinterred and cited. It was not, during the interim, lodged in an archive; most medical libraries almost certainly had copies of his books. It had been ignored, because it was unacceptable.

Ultimately, only the members of expert committees themselves can describe how they do produce estimates, and what factors cause their estimates of requirements to vary. None has (yet) begun to publish descriptions of their deliberations. Insofar as they do informally discuss their conclusions, it is clear that the impressive amendments of previous numerical values that they issue do not result from more and more precise determinations of reality but from changes in policy. There is also an important element of horsetrading, where X's support of Y's view of something is coupled to Y's advocacy of X's favourite value for something else: while the long-term impact of both X and Y in the haggling process may be truncated by Z, representing the sponsoring organization, who may not appoint them to the next committee.

4.   THE CONCEPT OF NUTRIENT REQUIREMENTS

Although coherent statements can be made about the nature of nutrient requirements, the phrase has been fitted to values based

on a variety of criteria, such as what an apparently healthy
group appear to eat or even what a committee think they should
eat.    Obviously, such estimates are not open to rational study.
The most general definition of a nutrient requirement is that
it is the minimal amount of nutrient needed to maintain a given
physiological state.    It seems clear, therefore, that before the
requirement can be specified, the physiological state which it is
desired to maintain must be specified in objective terms.    Further,
it will not be sufficient to do this in general descriptive terms
such as 'maintenance' without specifying what level is to be maintained, nor 'health', since we have no way of describing this
except as the absence of any conceivable type of disfunction.


## 5.    REQUIREMENTS FOR HEALTH

It is widely assumed that nutrient requirements are related to
health (or nutritional status), but in practice the utility of
health as a reference criterion is limited by our inability to define a state of ideal health which the adequate nutrient intake
should sustain, and by our consequent inability to detect small
deviations from it.

The social nature of the definition of health has been frequently
discussed elsewhere (see, for example, Illich (1976) or McKeown
(1979)), and need not be reviewed here.    Where the index of
normal function used is not even morbidity, but an acceptable
level of some physiological parameter such as an enzyme activity,
the arbitrary nature of the 'normal' state is even more apparent.

The use of health as a criterion for fixing energy requirement
is particularly complicated by the fact that with regard to
nutrients it seems impossible to identify a healthy population.
As FAO/WHO (1973) state:

"Most of the recommendations on energy requirements in this
report  are based on measurements of food intake...   If food
intake is used as the criterion of requirements certain assumptions are made.    One is that the population are healthy and this
may not be quite true.    Many groups of people... living in the
industrialized countries are obese.    On the other hand some
groups living in developing countries are small in stature, light
and thin yet may not be physically less healthy because of their
different body size".

One need only add that in some important respects they may well
be fitter and the whole basis for the 'requirements' is undermined.

6.   NUTRIENT BALANCE

    Nutrient balance is the second type of criterion used to define
energy requirements, and indeed requirements for many other
nutrients.   It is a deceptively simple notion which has found
wide applicability in the attempts to define nutrient needs for
animal production (see, for example, Blaxter 1967).   The balance
method ignores health, and is concerned only with the relationship between nutrient intake and expenditure.
    Obviously, in the non-pregnant adult animal, intake and ex-
penditure must match if the body content of nutrient is not to
rise or be depleted.   The problem is that balance can be achieved
over a range of intakes through adaptations of various kinds.
The question 'which level of equilibrium is preferred' can only
be answered by referring back to 'health'.
    The balance method is hence no more objective than the use of
health.   The decision about what level of equilibrium or, for
children, what rate of growth, and hence what degree of positive
balance, will be regarded as a norm, remains entirely subjective,
unless we are prepared to specify a particular set of desired
functions and a set of undesirable symptoms we wish to avoid.
    The decision which is usually made (albeit not always explicitly)
is to say that we prefer the levels and growth rates which are
typical of Western developed communities.


7.   TECHNICAL LIMITATIONS OF ENERGY REQUIREMENTS

    Since we have such strong reservations about the political
and structural constraints on expert committee reports, it might
seem strange, now, to undertake a technical critique of such a
report, ipso facto taking the report at its own valuation.
    But to avoid doing so would be to ignore the culturally based
power that tables of numbers exert, and to imply that, if only
previous criticisms can be met, a useful expert committee report
will result.   Because the 1971 FAO/WHO committee is the most
recent to report, and because its report (FAO/WHO 1973) is so
comprehensive, we shall use this as a case study for this critique.
    The energy requirement estimates proposed by the FAO/WHO
committee are based on food-intake data of apparently healthy
Western populations.   Details of the adults studied are not given.
Data for children are based on subjects 'growing normally'; 
data for infants aged less than six months are based on breastfed
children, growing 'normally' (FAO/WHO 1973, p.33).
    The committee's own reservations about the health of these

Western populations have already been noted. Apart from this brief cautionary note however, they set aside the problem and apparently seek to validate their estimates by using the criterion of energy balance, which they do by a factorial analysis of energy expenditure.

The study of balance at a single level of intake can demonstrate the adequacy of that intake, but its use to define a requirement implicitly rejects the adaptive model and accepts the notion that man is a fixed efficiency machine.

Some of the 19th century physiologists cited in table 2 made this assumption about man and were therefore willing to define requirement by measuring intake. Others - notably Benedict (1907, 1915, 1919) - seemed more aware of feedback mechanisms, and produced impressive evidence that energy expenditure is reduced when intake falls, and that balance might be achieved over a range of intakes. These observations would provide a rational explanation for the frequent reports that populations exist in 'developing countries' who habitually consume extremely low food intakes without catastrophic results (Miller & Rivers 1972, Norgan et al. 1974).

However, both this metabolic and epidemiological evidence is ignored by the FAO/WHO committee, as by all others. The reason is unstated, but appears to be that studies on energy expenditure in 'underfed' people are dismissed because being 'underfed' is pathological. What, if the possibility of adaptation to low food intakes is being seriously considered, is meant by 'underfed' is not clear, unless it is a statement that if you eat less than the experimenter or expert, you are undernourished.

This curious tautology may become less tenable with the acceptance of evidence that dietary-induced thermogenesis exists. Subjects who increase their energy intake in experimental studies are also observed to increase their energy expenditure to an extent that cannot be explained by variations in activity level (Miller 1975).

Adaptation in the underfed might be dismissed as pathological, but adaptation in the overfed will be accepted. And once it is accepted that man (like other animals) might be capable of maintaining zero energy balance over a wide range of intakes, then intake studies cease to be a valid way of defining a minimal requirement.

Moreover, their conjunction with simultaneous studies of energy expenditure does not improve the quality of the evidence. At best, it provides an extremely tedious and inaccurate confirmation of the first law of thermodynamics. Only estimates of minimal rates of expenditure are of any use. Thus, if the use of factorial analysis of expenditure by the FAO/WHO committee is to be seen as anything more than a version of the three card trick, then they presumably are presenting their data as true estimates of obligatory losses and maximal efficiencies.

The process of factorial analysis in defining requirements, due to Wood & Capstick (1928), partitions the energy expenditure of an animal into three separate compartments: maintenance, activity and growth. These are assumed not to interact, and it is commonly further assumed that the maintenance requirement can either be estimated directly, or from a study of minimal rates of energy expenditure (obligatory or endogenous losses) and the efficiency with which dietary energy will replace these losses.

The FAO/WHO committee use basal metabolic rates (BMR) as their estimate of minimal rates of energy expenditure. BMR is an arbitrary measurement defined, for the convenience of clinicians, as the metabolic rate of an individual, resting, comfortable and warm, after an overnight fast of 12—15 hours. It is often not even the minimal metabolic rate of that individual – metabolic rates below BMR are frequently recorded during sleep (Benedict 1915). Moreover, prolonged total or partial fasting reduces BMR values considerably, even when corrected for weight or estimated surface area, as table 4 shows.

The FAO/WHO committee make use of a standard set of BMR values collected by Talbot (1938) on 2200 healthy (presumably non-malnourished) citizens of Boston. Table 4 suggests that for any underfed population, values between 10% and 30% below these standard BMR values might be a better estimate of obligatory losses. Interestingly, and perhaps coincidentally, in Western populations the lower limit of the normal range of BMR (corrected for age, sex and weight) is also estimated at 20% below the average normal value.

Direct measurements of the efficiency of utilization of metabolizable energy for maintenance in man are scarce. The FAO/WHO expert committee approached this problem by an ingenious comparative approach which provided a reliable estimate. Repeated studies have shown that BMR in adult members of a wide range of species of mammals varies exponentially with weight and Kleiber's (1961) proposed value of 70 kcal per day per kg body weight to the power 0.75 ($W^{0.75}$) is now generally accepted.

The FAO/WHO committee collated the evidence on energy intakes for zero energy or zero nitrogen balance in experiments on animals and man and showed that this too was related to $W^{0.75}$. The mean value of 105 $W^{0.75}$ kcal/day, which has been confirmed by various subsequent studies, encouraged the committee to propose that 1.5 x BMR should be taken as the minimal energy cost of weight maintenance, and this value is used in their factorial analysis of energy expenditure.

The problem of whether this ratio is still applicable to individuals in whom BMR is reduced by food restriction can be

TABLE 4. The effect of undernutrition on BMR. Mean values reported from various studies.

| | Number of Subjects | Days of fast | Mean intake during fast (kcal/day) | % Change in BMR | | |
|---|---|---|---|---|---|---|
| | | | | Per person | Per unit weight | Per unit surface area |
| **Total starvation** | | | | | | |
| Benedict (1915) | 1 | 31 | nil | 29 | 10 | 15 |
| Kleitman (1926) | 1 | 40 | nil | 36 | 19 | 18 |
| **Partial starvation** | | | | | | |
| Benedict (1919) | 24 | 120 | 1400 | 19 | 12 | 16 |
| Berkman (1930) | 117 | – | Anorectics | – | – | 25 |
| Beattie & Herbert (1947) (a) | 11 | >100 | <1750 | 26 | 14 | 16 |
| Keys et al. (1950) | 33 | 200 | 1570 | 39 | 19 | 31 |
| Grande (1964) (1) | 13 | 20 | 1000 | 17 | 9 | 13 |
| (2) | 12 | 14 | 1000 | 21 | 14 | 18 |

(a) BMR per unit surface area provides the best approximation to the relationship between BMR and weight in Talbot's standards. The changes reported in this row are probably therefore the most applicable in the context of the factorial calculation of energy requirements by WHO/FAO (1973).

answered in part by consideration of Benedict's (1915) and Keys
et al's (1950) studies of semi-starvation.  Both groups of subjects
were inactive and virtually maintaining weight at the end of the
experiment.   Their BMRs had fallen, but the mean ratio of estimated maintenance intake to BMR was 1.5 for Benedict's subjects
and 1.6 for Keys'.    Beattie & Herbert (1947a,b,) and Beattie
et al. (1947) produced data on adults recovering from severe
malnutrition which give a value of 1.4 BMR for maintenance. It
is probable, therefore, that the value 1.5 can be applied to
populations on low food intakes with some confidence.
     The two other components of the factorial analysis are the
energy costs of growth and activity.  The energy costs of growth
have been a subject of great interest amongst both animal and
human nutritionists, but essentially because of the scientific interest of the subject rather than, as is claimed, its importance
in helping determine energy requirements.
     The FAO/WHO committee took from the literature values of the
energy cost of protein and fat deposition and, assuming that
weight gain is 18% protein, 16% fat and 66% water, calculated a
cost of 5 kcal above maintenance per g tissue gained.   These
values, which assume 16 kcal/g protein retained and 13 kcal/g
fat retained, are well within the range of estimates that can be
found in the literature from animal experiments that are based
on a variety of techniques (FAO/WHO 1973).
     Studies relating weight gain to intake in recovering malnourished
children yield higher values (values of 10 kcal/g gain have been
cited) (Ashworth 1969).    Stoichiometric calculations yield lower
values (2 kcal/g) and some recent experiments in pigs (Reeds
et al. 1981) confirm Mitchell's (1962) suggestion that the energy
cost of growth might in fact vary with the composition of the diet.
     However, given the apparent errors that might be involved in
other parts of the requirement estimate, the accuracy of the
contribution due to growth matters very little:  human growth
is so slow that, after infancy, its energy cost accounts for
less than 2% of the total energy requirement.
     What matters much more when the requirements are applied to
children is the body weight that is assumed.  The FAO/WHO expert
committee recommend that, in calculating the requirements of a
community, the requirement of children below 13 years should be
assessed by age, but not adjusted for actual weight, even when it
is known that the average weights are lower than 'normal'.  The
"...intention is to provide for 'catch up growth' and a return to
the 'normal' height and weight..." (FAO/WHO 1973 pp.33-34).
     Like so much else in the report, it is difficult to see if this
really means what it appears to say.   Certainly, its effects are
startling.  Since the bulk of the energy needs of children relates

to maintenance and activity and not to costs of growth, this procedure ensures that food shortages in populations containing underweight children are exaggerated beyond rehabilitation needs by exaggerating maintenance and activity costs.

Consider, for example, a child gaining weight at 70% of the reference rate employed by FAO/WHO. Between its 2nd and 5th birthday such a child will gain 4.5 kg as compared to 6.06 in the reference population. The 'energy requirement' suggested by the FAO/WHO will total 600 000 kcal during those three years since they will assume 'normal' rates of weight gain over the period. But this is 150 000 kcal more than that which would have resulted if requirement had been related to actual body size. In fact, the cost of catching-up on the basis of 5 kcal/g would be (6.06 - 4.5) x 5000, i.e., less than 10 000 calories. Thus, the committee's method of calculating the requirement suggests 15 times more extra energy is needed than the actual cost of catch-up growth.

This curious correction procedure with children guarantees that small populations will apparently be underfed. It may be they are. It may be that we should all grow at the Western rate and end up bemoaning the curse of the diseases of affluence. Or it may be that the whole procedure is pseudo-physiology, a method of manufacturing malnutrition, a way of classifying the small as the malnourished, just as the uncritical anthropometric comparison of children in developing countries with Western standard populations has long been used to achieve the same end.

The final compartment of energy expenditure considered by the FAO/WHO committee is activity. If populations with low food intakes adapt to these intakes, then the contribution of activity to such adaptation could come either through a decrease in energy cost of specific activities or through a reduction in total expenditure. It seems probable that both types of adaptation might occur – certainly, voluntary activity has been reported by Keys to decrease in experimental semi-starvation (Keys et al. 1950). Keys also reported that although the energy cost of specified tasks was the same per unit weight throughout semi-starvation the total energy cost of these tasks was reduced because body weight fell. Some physiologists have claimed that 'work capacity' assessed by measures such as maximal $\dot{V}O_2$ is also impaired by undernutrition (Keys et al. 1950). It also does not, in an 'undernourished' population, decline so severely with age as in an affluent one.

The term work capacity is anyway another misleading piece of professionalism, since the real amounts of work done are not dictated by $\dot{V}O_2$ max but by social factors.

A further limitation on the validity of the FAO/WHO estimates of requirement is the correction of the reference values promulgated by the committee for body weight. Scaling requirements for body weight is a general problem in estimation of energy requirements. Since the intakes of the reference population which are used to fix the requirement seem to be only poorly correlated with weight (Ferro-Luzzi; this volume, p. 115) the theoretical basis of any scaling process must be shaky.

The FAO/WHO report demonstrates this clearly. Table 3 in their report gives energy requirements for adult males in the weight range 50–80 kg. These are linearly related to body weight. However, the BMR data they use are, over the same weight range, proportional to $W^{0.56}$, and so, therefore, is the calculated cost of weight maintenance.

This leads to the strange situation where the cost of activity is therefore proportional to $W^7$ (it is linearly related to weight in all other species (Blaxter 1967)). It also suggests that at low body weight the cost of work is negative. If the report is taken seriously, we must conclude that light activity for an 80 kg man costs 450 kcal/day above maintenance, but for a 50 kg man light activity can be undertaken on an intake of 100 kcal <u>below</u> maintenance.


8.  CONCLUSIONS

We have discussed some of our reservations about current esti- mates of energy requirements, and the system which generates them, at considerable length. Such a broad-ranging criticism will inevitably be condemned as totally negative: but then it is difficult to adopt any other stance, in view of the unresolved problems in the subject which is being reviewed.

A measure of these can be gauged from FAO's use of its own report. The year after the FAO/WHO expert committee had met, FAO itself attempted in its 4th World Food Survey (FAO 1973) to measure the extent of malnutrition in the world by comparing Food Balance Sheet data with requirements. It was unable in doing this to make use of its own expert committee report on requirements. Faced with precisely the problem that what is needed is in fact a limit of adaptability, it arbitrarily took the lower limit of normal variation of BMR (80% of mean value) as a limit of adaptation. Hence, it adopted a value for the minimal cost of maintenance of 1.5 x 0.8 x BMR (= 1.2 BMR). This value – based on Talbot's 50 year old data – it took as the cut-off point for defining malnutrition. The energy requirements report was circumvented.

Perhaps no more severe condemnation of the outpourings of expert committees is possible than the fact that the sponsors themselves seem to find them unusable.

REFERENCES

Anon., 1952, The Economist (London), 23 August, 1952, p.456.
Ashworth, A., 1969, British Journal of Nutrition, 23, 835.
Atwater, W.O., 1895, Bulletin of the United States Department of Agriculture, 21, 46.
Beattie, J. & Herbert, P.H., 1947a, British Journal of Nutrition, 1, 183-192.
Beattie, J. & Herbert, P.H., 1947b, British Journal of Nutrition, 1, 192-202.
Beattie, J., Herbert, P.H. & Bell, D.J., 1947, British Journal of Nutrition, 1, 202-218.
Benedict, F.G., 1907, The Influence of Inanition of Metabolism (Washington D.C.: Carnegie Institute).
Benedict, F.G., 1915, A Study of Prolonged Fasting (Washington D.C.: Carnegie Institute).
Benedict, F.G., 1919, Human Vitality and Efficiency under a Prolonged Restricted Diet (Washington D.C.: Carnegie Institute).
Berkman, J.M., 1930, American Journal of Medical Science, 180, 411-424.
Blaxter, K.L. 1967, The Energy Metabolism of Ruminants, 2nd Ed., (London: J. Hutchinson Ltd).
British Medical Association (BMA), 1933, Report of the Committee on Nutrition of the BMA (London: British Medical Association).
British Medical Association, 1950, Report of the Committee on Nutrition of the BMA (London: British Medical Association).
Chittenden, R.H. 1905, Physiological Economy in Nutrition (London: W. Heinemann & Co.).
Chittenden, R.H., 1907, The Nutrition of Man (London: W. Heinemann & Co.).
Department of Health and Social Security, 1969, Recommended Intakes of Nutrients for the United Kingdom (London: HMSO).
Department of Health and Social Security, 1979, Recommended Daily Amounts of Food Energy and Nutrients for Groups of People in the United Kingdom (London: HMSO).
Food and Agricultural Organization (FAO), 1950, Calorie Requirements (Washington D.C.: FAO Nutrition Studies No. 5).
Food and Agricultural Organization (FAO), 1957, Calorie Requirements (Rome: FAO Nutrition Studies No. 15).
Food and Agricultural Organization/World Health Organization, 1973, Energy and Protein Requirements (Geneva: WHO Technical Report Series No. 522).
Gautier, A., 1907, L'Alimentation et les Régimes chez l'Homme Sain et chez les Malades (Paris).

Grande, F., 1964. In Handbook of Physiology: Section 4 -
Adaptation to Environment, edited by D.B. Dill (American
Physiological Society, 1964), pp.911-937.
Hultegren, D. & Landegren, O., 1891, Untersuchungen Über
Die Ernährung Schwedischer Arbeiter (Stockholm).
Hutchinson, R., 1906, Food and the Principles of Dietetics,
lst Ed. (London: Edward Arnold).
Illich, I., 1976, Limits to Medicine. Medical Nemesis: The
Expropriation of Health (London: Marion Boyars).
Indian Council for Medical Research (ICMR), 1944, Dietary
Allowances for Indians (Hyderabad: ICMR).
Indian Council for Medical Research (ICMR), 1960, Dietary
Allowances for Indians, Revised Ed. (Hyderabad: ICMR).
Indian Council for Medical Research (ICMR), 1968, Dietary
Allowances for Indians, revised Ed. (Hyderabad: ICMR).
Indian Council for Medical Research (ICMR), 1971, Dietary
Allowances for Indians, Revised Ed. (Hyderabad: ICMR).
International Union of Nutritional Sciences (IUNS), 1975, Report
of the Committee on International Dietary Allowances. Nutrition
Abstracts and Reviews, 45, 90-111.
Keys, A., Brozeck, J., Henshel, A., Mickelson, O. & Taylor,
H.L., 1950, The Biology of Human Starvation (Minneapolis:
University of Minnesota Press), two volumes.
Kleiber, M., 1961, The Fire of Life, lst Ed. (New York: John
Wiley).
Kleitman, N., 1926, American Journal of Physiology, 77, 233-244.
Kuhn, T.S., 1970, The Structure of Scientific Revolutions, 2nd
Ed. (Chicago: University of Chicago Press).
League of Nations Health Organization, 1935, Report on the Phy-
siological Bases of Nutrition (Geneva: League of Nations
Publication Series No. 3).
Lichtentfelt, H., 1903, Pflügers Archiv für die gesamte Physio-
logie der Menschen und der Tiere, 99, 1.
Lusk, G., 1909, The Elements of the Science of Nutrition, 2nd
Ed. (Philadelphia and London: W.B. Saunders Co.).
Miller, D.S., 1975, The Regulation of Energy Balance in Man
(Geneva: Medicine & Hygiene).
Miller, D.S. & Rivers, J.P.W., 1972, Proceedings of the Nutri-
tion Society, 31, 32A.
Mitchell, H.H., 1962, The Comparative Nutrition of Man and
Domestic Animals (New York: Academic Press), two volumes.
Moleschott, J., 1859, Physiologie der Nährungsmittel: ein Hand-
buch der Diatetik (Giessen: Farber).
National Research Council, 1943, Recommended Dietary Allowances,
lst Ed. (Washington D.C.: National Academy of Sciences).
National Research Council, 1945, Recommended Dietary Allowances,
2nd Ed. (Washington D.C.: National Academy of Sciences).

National Research Council, 1948, Recommended Dietary Allowances, 3rd Ed. (Washington D.C.: National Academy of Sciences).
National Research Council, 1958, Recommended Dietary Allowances, 5th Ed. (Washington D.C.: National Academy of Sciences).
National Research Council, 1964, Recommended Dietary Allowances, 6th Ed. (Washington D.C.: National Academy of Sciences).
National Research Council, 1968, Recommended Dietary Allowances, 7th Ed. (Washington D.C.: National Academy of Sciences).
National Research Council, 1974, Recommended Dietary Allowances, 8th Ed. (Washington D.C.: National Academy of Sciences).
Norgan, N.G., Ferro-Luzzi, A. & Durnin, J.V.G.A., 1974, Philosophical Transactions of the Royal Society of London, Series B, 168, 309.
Poleman, T.T., 1977, World Development, 5, 383-394.
Price, D.J. De Solla, 1963, Little Science Big Science, (New York: Columbia University Press).
Ranke, J., 1876, Grundzüge der Physiologie des Menschen (Leipzig: W. Englemann).
Reeds, P.J., Fuller, M.F., Cadenhead, A., Lobley, G.E. & McDonald, J.D., 1981, British Journal of Nutrition, 45, 539-546.
Schmidt, A., 1901, Die Militärarzt Zeitschrift, p.622, cited from Hirchfeld 1903.
Studemund, J., 1878, Pflügers Archiv für die gesamte Physiologie des Menschen und der Tiere, 48, 578.
Talbot, F.B., 1938, American Journal of Diseases of Childhood, 55, 455.
Voit, C., 1881, Handbuch der Physiologie, edited by L. Herman (Leipzig), p. 519.
Wood, T.B. & Capstick, J.W., 1928, Journal of Agricultural Science, 18, 486.
Wretlind, A. (Moderator) 1977, A Round Table on Comparison of Dietary Recommendations in Different European Countries. In Proceedings of the Second European Nutrition Conference, edited by N. Zollner, G. Wolfram and G. Kellner (Basel: Karger), pp.185-255.

# DETERMINANTS OF NUTRITIONAL NEED

H. Weymes
Department of Anthropology,
University College,
Gower Street, London WC1

Of all the questions facing nutritionists and human biologists today, that of defining nutritional need is one of the most important. This is because a definition of such need is required, for the planning of food supplies, by economists who must estimate the need to import foods, by planners of agricultural development and of the food-manufacturing industry, and by the infrastructures needed to support these. Also, some provision must be made for possible shortfalls in food supplies through natural disasters or social upheavals. Someone has to assess how much food is required to feed populations.

An assessment based on the observed food-intake levels of healthy people would be useful, if it were known what the meaning of 'health' is in this context. Some of the diseases of affluence are believed to have their beginning in what might appear to be very healthy diets some 20 years before the onset of the disease, so the intake study would have to be very long-term before a diet would be accepted as 'health-promoting'. The use of animal models for shorter-term experimentation with diets provides direct answers to questions about the physiology of nutrient utilization in the animal body, but studies on human populations are necessary if we are to understand the needs of humans within their social settings. It is not appropriate to use only laboratory studies to define human needs when, as can be shown, such needs are closely related to emotional needs and may vary with environmental hazards. In most instances, it is difficult to unravel the complex interaction between these factors. The first natural experiment in this area was described by Widdowson (1951) who set out to discover the effect of increased rations on children in German orphanages. One orphanage was given a food supplement after a six-month period of control. A second orphanage was not given the supplement, but was also monitored during the whole period of the experiment. The children in the first orphanage gained more weight than those in the second in the period of control when no supplement was

given, and then surprisingly, gained less weight during the
subsequent period of supplementation.    The explanation rested
on the transfer of the matron from the second orphanage to the
first at the time supplementation began.    A domineering, strict
woman, she upset the children by upbraiding them at mealtimes,
upsetting them emotionally.

How should nutritional need be defined?    This question
should be dealt with on several different levels.    Firstly, there
is the need of people who are temporarily distressed by some
environmental disaster or by a social upheaval.    For these, the
need is to be kept alive and well until some solution can be found
for their predicament.    This problem lies with social-welfare
workers and with politicians.    Internationally defined standards
of nutritional recommendations are useful for such short-term
events, but are not necessarily appropriate for longer-term requirements.    For the great majority of the world's population,
however, the level of nutritional need is one that will not only
sustain life but will allow the satisfaction of some choice from a
range of foods, in amounts that will give people the freedom to
share food in excess of bare requirements with friends and relations.    Food is central to the social life of everyone, under
normal circumstances, and is part of their cultural experience
rather than the means of keeping body and soul together.
Another level of food need is for the quantity and variety of
foods that will promote maximal growth at an early age.    This
has certainly been the level believed desirable until recently,
but now there are second thoughts about the value of fast
growth and large size.    Once it is understood that there is
more than one level of nutritional need, it is easier to handle
the question of 'how much', by asking, 'for whom', and
'for how long'?    Not only must it be clear what the target population is, it is also important to try to assess the adaptations
shown by the population in question to their environment.

The majority of individuals are capable of adapting to great
variations in such environmental conditions as changes in temperature, humidity and altitude, and to a wide range of nutrient
intakes, both in food types and in the quantity consumed.
Customary food-intake levels that are quite adequate for an ac-
climatized population may be described as inadequate or as
excessive where an inappropriate level of requirement is used
as a standard.    Intake levels that are satisfactory for people
who are largely self-sufficient in food production, and who are
occupied with farming and home-making, may be quite inadequate
for the same people under the stress of unemployment or displacement.    The social setting is of great importance and this
aspect of nutritional 'needs' requires further study.    Levels
of nutrient intake that are satisfactory when derived from a

variety of foods may not be so adequate when supplied from
a restricted range of foods, because the boredom of a mono-
tonous diet acts as a suppressor to appetite, lessening the a-
mount of food eaten.    So if we can define a variety of dietary
patterns and describe a range of food-intake levels in different
social contexts where people show no sign of malnutrition, we
may be able to describe the intake levels that are maximal as
indicators of need in terms of both social and nutrient content.
(Here, malnutrition should be interpreted in terms of morbidity
and mortality from the common diseases of under- and over-
nutrition:   failure to grow adequately in childhood, a raised
under-five death rate, poor immune response to infection, and
diseases of the cardio-vascular system come to mind).    The
lower intake levels would be useful as realistic guides to the
real food requirements of many communities today.

Environmental health is so important for nutritional success
that the basic facts about clean-water supplies and levels of in-
fectious disease are part of the overall pattern of food needs.
Food that cannot be absorbed because of intestinal infection is
food wasted, and the anorexia of fever depresses the availability
of raw materials for the use of the immune system.    There is
some evidence of an adaptive response to some infections, however.    Malaria and measles are both more severe in a community
that is newly exposed to them, while communities where these
have been endemic for some time are less severely affected.
It has also been suggested that in some instances poor nutrition
may be responsible for a low level of infection, since the parasites are kept in check by the poor availability of nutrients,
but it is not clear whether or not this is universally the case.
A report on the Turkana herders of the shores of Lake Rudolph
describes the level of infection in two groups, one of which ate
fish and milk, while the other more traditional group lived on
milk alone.    Both groups were equally tall and heavy, but the
fish eaters suffered appreciably more from infections and worm
infestations than did the milk drinkers.    Malaria, hookworm
and amoebas were all more common in the former group, the
suggestion being that the improved haemoglobin levels in the
blood of this group allowed the parasites to flourish more readily
(Murray 1980).

It is clear from dietary surveys from many communities that
some deficiencies are due in large part to lack of information
about nutritional requirements, for there are frequently sources
of vitamins and minerals in plants that are available and inexpensive, but that are largely ignored by the deficient population.
The most commonly reported deficiencies of this sort are of
calcium, vitamins A and C, and iron.    In many studies, the real
intake of these nutrients is under-reported because the method

of food preparation isn't well understood.    In the preparation
of cereals, the contribution of the lime used in the processing
is ignored, and fruits and berries are so commonly outside the
classification of 'food' in the vocabulary of many people that
they are unwittingly passed over in dietary recall records.
Recently, an anthropologist sent in a list of 14 named varieties
of wild-plant foods that he had observed being consumed throughout the year by members of a community whose intake of vitamins
A and C was recorded as virtually nil in the official survey of
dietary intake.    Such surveys often use enumerators from another area or from a sector of the community not thoroughly ac-
quainted with local nutritional habits.    In cases like this, where
no signs of deficiency are seen, yet the survey shows a deficiency
of intake by the usual method of recalling foods eaten, or by a
weighed intake survey lasting a few days, the question must be
asked:    'how realistic are these intake surveys in fact, and should
nutrition intervention be planned on the basis of this level of information?'    It is argued that deficiency may be 'subclinical',
whatever that means, and that the general health of the people
will improve with improvements to the diet.    That seems to me
to be true if by improvements the intention is to increase the
quantity and variety of foods within reach of the people and to
give them the information about their dietary deficiencies, and
about the value of the improvements being offered.    But if it
means launching feeding programmes or fortifying everyday foodstuffs with the items deemed 'deficient', there is a lot to be said
for trying to find out what the real food intake is by using ob-
servant fieldworkers rather than relying on the survey data alone.
Where no deficiency diseases can be detected but the child population is small and light for age, it is probably more constructive
to try to improve the level of intake of available foods than to
try to introduce new varieties, or feeding programmes, or even
undertake a detailed food-intake survey.    These can all become
substitutes for actually doing anything to increase the availability
of basic foods to the community.    Food fortification programmes
are rather fashionable at present, and do pose something of a
dilemma in some situations.    It is perhaps easier to legislate for
everyone's sugar or tea to be fortified with vitamin A than to
get fresh fruit and vegetables to any target group.    However,
the chances of anyone trying to improve the diet of an impoverishe
population, once out of danger from a serious deficiency, are
pretty slim.    Taken to extremes, it should be possible in the
future to fortify two or three basic ingredients with most of the
nutrients required by man, and to keep people 'healthy' on a
diet replete with nutrients and totally lacking in variety and
stimulation.

Sometimes the restriction of diet is associated with the separation of one segment of a community on migration, either to find work or due to a resettlement programme.   Lack of knowledge of locally grown foods may result in a reliance on store-bought goods, even when land is available for gardens and opportunities exist for keeping chickens and goats.   A study by Desai et al. (1980) based on an agricultural community in Brazil, recorded malnutrition in the presence of an abundant supply of fruits and edible leaves growing wild, and ample time and space for gardening and keeping livestock.   The foods purchased with the wages earned by the men from agricultural labouring were mainly alcoholic and sugary drinks, white flour and white sugar. The total lack of interest in producing a home-grown supply of food appears to be associated with the separation of this group of people from the parental generation or perhaps with a lack of social contacts with small farmers in the area.   This sort of a community is in need of an education programme, perhaps using community-based people to demonstrate possible alternatives to store-bought foods, rather than an authoritarian approach.

Groups that restrict the range of foods in their diet from choice, not ignorance of their availability, include the Bushmen. Lee (1968) reported that only 23 of the 85 known species of edible plants were used and 17 of the 54 known species of animals eaten.   Perhaps both Brazilians and Bushmen can be said to be saving energy by restricting their dietary range, but in the case of the Bushmen they are still eating a wide range of foods; they are not malnourished, and they choose to do other things in their spare time, like visiting friends and playing games and being sociable.   In contrast, the Brazilians' social activities seem to have been as meagre as their diets, possibly because of a loss of community identity following their migration.

Food restriction may be imposed by religious beliefs.   This is most vividly illustrated by the study of Levine (1965), in which he describes the fasting and feasting in Christian Ethiopia.   He says that in this community "the ordinary Christian is obliged to fast on 165 days each year and the devout faithful approximately 250 days".   The effect of the fasting, which lasts the whole day and not just until sunset like the Muslim fasts, is widespread malnutrition.   This restriction is voluntary, not due to ignorance in the way the Brazilians were ignorant, nor to choice amongst plenty as with the Bushmen.   Perhaps an understanding of the association between fasting and child malnutrition is lacking, perhaps not. Only someone with a thorough understanding of the value of health in that society could give an opinion on the acceptability of any intervention.

Real poverty in dietary terms lies in a lack of choice.   Many people subsist on diets that are nutritionally adequate but so restricted in their components that the monotony very quickly quenches the appetite and food intake is reduced accordingly. The needs of children are very important in this respect.Children often pass through periods of mild ill-health when they have to be tempted to eat enough for their requirements for growth. Inexpensive ways of spreading information about the value of variety and of using fruits and vegetable shouldn't be too difficult to devise in these days of the ubiquitous radio. There can be no excuse for neglecting to spread information about dietary needs to people who have a choice in the food they buy or in the foods they grow and gather. Education is generally unpopular because the results are so slow in coming that the initial input is usually forgotten and the effort designated a failure. That people's dietary habits can and do change is demonstrated widely, but emphasis is given to deteriorating diets because these give cause for concern. Where new foods are adopted through choice, it is obviously not going to be easy to get people to change back to some other foodstuff, but where the change is from necessity, a reversal or change to another foodstuff should be relatively easy to accomplish. Thus, the motive for dietary restriction or change must be understood before a dietary improvement is suggested.

The composition of human diets is extremely diverse and may consist entirely of plant foods, of animal or fish foods, or of mixtures of the two. But the majority of people manage on a few staple roots and cereals with seasonally variable amounts of animal produce, some fruit, and vegetable garnishes. The range of total food-intake levels in terms of energy and protein is also very great. The energy level is of supreme importance in that most foods contain some protein but it can only be used for anabolic purposes when energy needs have been satisfied. Clearly, there are limits to both high and low levels of energy intake, but it is becoming apparent that individuals differ considerably in the efficiency with which they convert energy into tissue stores. The WHO recommended levels are designed to meet the energy needs of the majority of all people, but these are often treated as minimal for survival, which clearly they are not. Many communities are reported to have energy-intake levels at around two-thirds of the recommended levels without showing signs of malnourishment, and other communities have intakes one-third to one-half higher than the recommended levels, with signs of obesity not necessarily amongst those who eat the most.

Why some individuals are protected from the effects of undereating or overeating when others are not is unknown at present. What is of interest is that, although the food intake may not be that recommended, if it is not the cause of any disability or ill-

ness, it is not necessarily harmful. The non-food factors associated with the lifestyle of healthy, active people should be
investigated, because it is these that are the clues to what pre-
cipitates nutritional disease and malnutrition. What converts a
community on a low-energy intake into a malnourished community?
Factors that are obviously implicated include environmental sani-
tation, housing conditions and education. But a really important
(because unknown) factor is that of motivation. Why do some
people make the most of a difficult environment, of upheaval and
the stress of forced migration, when others find these problems
overwhelmingly difficult to overcome? Why will some people work
hard to tend gardens and cook imaginative meals from poor foodstuffs, when others prefer not to bother to prepare tempting
meals? Is it because of cultural differences, differences in food
appreciation learned in childhood, and if so can these be overcome?
We know that they can, under some circumstances, from our own
food habits, which are not those of our grandparents, still less
of our great-grandparents. Certainly, it seems that newly re-
settled people in both urban and rural environments need some
advice in choosing a new diet, and the selection of a balanced
diet obviously isn't instinctive but has to be learned. Where
poor diets are selected through ignorance, some guidance is the
responsibility of the informed sector of the society, and an understanding of methods of food preparation is needed if a change in
diet is advocated.

It seems that, at present, household food-intake surveys alone
are not sufficiently accurate where there are marginal deficiences
of vitamins or minerals. This is due to several factors, which
include the difficulty of assessing food intake accurately without
disrupting the eating habits of the population being studied, the
assessment of the nutrient content of the actual foods being eaten,
which may vary with the area in which they are grown and the
way they are prepared, and the problem of individuals' variability
in the ability to absorb the available nutrients as well as variation
in the individuals' needs for nutrients. These, and other difficulties, like the problem of food distribution within the family,
make the household food-intake survey method an unsuitable
basis for a planned intervention, like supplementary feeding programmes etc. The most commonly used measure of nutritional
adequacy is the growth rate of children, chiefly using height and
weight for age. This method has the advantage of being easy,
because nobody objects to being weighed and measured, and it is
cheap. Unfortunately, results are usually used to compare the
average heights and weights for age with those of a standard
population of North America (Boston). This would not matter
if it weren't for the fact that the smaller, lighter children of
rural tropical countries are sometimes described as 'malnourished'

simply because they are below the standard set by a group of children from a different stock, in a colder climate and with a more urban lifestyle. Children in rural areas have more free- dom of movement than those in urban surroundings and usually have obligations to help their parents on the land, fetching water and wood, or simply running errands. Children in cold, urban environments are often very restricted in their movements during their early years, and perhaps this may contribute something to their larger size. In any event, provided that children are free from disease in either environment, whether or not the genetic potential for height is reached seems to be of minor importance compared to the benefits of health. Independent standards of height and weight for age are being adopted now in many coun- tries and their use should be encouraged.

If present trends in population growth continue, our attitude towards nutritional needs must become more pragmatic and resources must be used for the prevention of nutritional deficiencies that are real. This means that prevention of the diseases we know to be most common, like vitamin A dificiency, should be antici- pated, and an education programme should be aimed at informing people of the possible consequences of neglecting to choose a varied diet. For those unfortunate people who have no choice, the responsibility for planning lies with their governments, in whose hands such alternatives as fortification and redistribution of food or wealth must rest.

REFERENCES

Desai, I.D., Tavares, M.L.G., Dutra de Oliveira, B.S., Douglas, A., Duarte, F.A.M. & Dutra de Oliveria, J.E., 1980, Food habits and nutritional status of agricultural migrant workers in Southern Brazil.   The American Journal of Clinical Nutrition, 33, 702-714.

Lee, R.B., 1968, What Hunters Do for a Living.   In Man the Hunter, edited by R.B. Lee and I. deVore (Aldine Publishing Co.), p.30.

Levine, D.N., 1965, Wax and Gold: Tradition and Innovation in Ethiopian Culture (Chicago: University of Chicago Press), p.232.

Murray, M.J., Murray, A.B. & Murray, C.J., 1980, An ecological interdependence of diet and disease?   American Journal of Clini- cal Nutrition, 33, 697-701.

Widdowson, E.M., 1951, Mental contentment and physical growth. Lancet, 1, 1316-1318.

# MEANING AND CONSTRAINTS OF ENERGY-INTAKE STUDIES IN FREE-LIVING POPULATIONS

A. Ferro-Luzzi
National Institute of Nutrition,
Via Ardeatina 546,
Rome, Italy

## 1.    INTRODUCTION

Energy-intake studies are obviously synonymous with food-intake measurement, as man obtains his energy from the chemical energy present in foods.

It is unusual for energy-intake data of free-living populations to be collected for the exclusive purpose of studying energy exchange and metabolism.    Energy-intake data are mostly generated, sometimes as by-products, in the frame of more general surveys designed to collect information on other nutritionally relevant aspects.

Dietary-intake studies are usually performed for assessing the nutritional risk of individuals or populations, or for obtaining information concerning uses, demand and disappearance of foods.    Less common aim is to achieve a better understanding of the socio-cultural determinants of eating behaviour.

The results of dietary surveys are used in the design and implementation of health policy regulations, for example, mineral or vitamin enrichment of staples to combat nutrient-deficiency syndromes, stipulation of maximum admissible levels of undesirable substances in foods to prevent toxic effects or cancer risk to populations, and identification of associations between dietary patterns and the epidemiology of chronic nontransmittable diseases or conditions such as atherosclerosis, cancer, hypertension, diabetes, obesity and liver cirrhosis.    Finally, dietary survey data may be useful for national planning for the production and distribution of food commodities.

These introductory remarks serve the purpose of indicating the central position occupied by food-intake studies in nutritional and epidemiological research and of identifying dietary survey methodology as a specific, delicate and highly flexible ad hoc instrument.    What I wish to emphasize is that, as for any instrument, this one also requires expertise and skill.    The operator has to be aware that a well defined

*Energy and Effort*

optimum range of measurement and an inherent measurement error are associated with the use of the instrument, that it needs calibration and must be standardized for different uses under different conditions and, finally, that each of the various methodological variants have their own appropriate prescriptions. Unfortunately, it seems that all this is rarely taken into consideration and that a variable degree of improvisation, justified on the ground of expediency, dominates the scene (Durnin & Ferro-Luzzi 1982). This, as well as faulty use, mishandling, or forcing of the capacity of the instrument inevitably results, as for any other instrument, in data which are difficult to interpret, are open to criticism, or are simply incapable of yielding results precise or reliable enough to provide the answer to the question addressed.

Although a number of papers have already appeared, arguing for the need for a more rigorous methodological approach and indicating the means of achieving it (Marr 1971, Lechtig et al. 1976, Garn et al. 1978, Liu et al. 1978, Beaton et al. 1979, Keys 1979), I feel it appropriate to devote the first part of this paper to discuss some methodological aspects, focusing on the collection of energy-intake data. The second part will be devoted to the illustration of some of the major problems currently associated with the interpretation of the results of dietary surveys.


2.   METHODOLOGICAL ASPECTS

The material to illustrate my points is drawn from several surveys carried out at different times, under different field conditions, and includes groups widely varying in size, age, sex, occupational status and ecological background. Such a variety qualifies them as particularly appropriate for the exemplification of the various aspects that will be discussed in this presentation. A summary of the relevant basic data of the selected food-intake studies is shown in table 1. All the surveys were performed with the precise weighing method, except for New Guinea, where the inventory method was used. Details of the exact procedures and of the characteristics of the groups have been described (Norgan et al. 1974, Ferro-Luzzi et al. 1975, Norgan & Ferro-Luzzi 1978, Ferro-Luzzi et al. in preparation, Ferro-Luzzi et al., unpublished data).

A variety of techniques is available for measuring food intake in free-living populations (see figure 1). They can be suitably divided into techniques which are appropriate for measuring food intake at the level of the household, and those

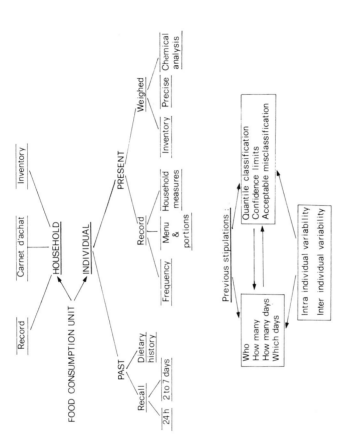

FIGURE 1.    Summary of the various techniques available for the assessment of food intake.
At the bottom of the figure, enclosed in three boxes, are the questions that
have to be addressed before beginning the fieldwork.

*Energy and Effort*

TABLE 1. Characteristics and mean daily energy intakes of various groups of adults and children.

| | Sex | No. | Age (y) | Body weight (kg) | Energy intake[1] (kcal/d) | Reference |
|---|---|---|---|---|---|---|
| **Adults** | | | | | | |
| Shipyard Workers, Italy | m | 150 | 37 | 76.4 | 3242 ± 50 | Norgan et al.1978 |
| Farmers, Italy | m | 24 | 43 | 74.3 | 3700 ± 116 | Ferro-Luzzi et al., in preparation. |
| Farmers, Finland | m | 29 | 42 | 81.4 | 3298 ± 77 | Ferro-Luzzi et al., in preparation. |
| Farmers, USA | m | 21 | 44 | 83.9 | 3327 ± 111 | Ferro-Luzzi et al., in preparation. |
| Peasants, New Guinea | m | 94 | 31 | 56.7 | 2209 ± 50 | Norgan et al. 1974 |
| Peasants, New Guinea | f | 110 | 29 | 48.6 | 1678 ± 42 | Norgan et al. 1974 |
| **Children** | | | | | | |
| Rural, Italy | m,f | 36 | 1–5 | 15.9 | 1243 ± 49 | Ferro-Luzzi et al., unpublished. |
| New Guinea | m,f | 110 | 3–5 | 14.2 | 1315 ± 37 | Ferro-Luzzi et al. 1975. |
| New Guinea | m,f | 157 | 6–10 | 20.8 | 1512 ± 31 | Ferro-Luzzi et al. 1975. |
| New Guinea | m,f | 167 | 11–17 | 34.5 | 1738 ± 39 | Ferro-Luzzi et al. 1975. |

[1]Mean ± S.E.M.

which attempt to measure the individual's diet.     Both have
their own limitations and spheres of validity.     However, the
concern of this paper will be only with the techniques which
attempt to measure the individual intake.     These have been
divided as is current practice, between measures of the present and of the past intake.

It is beyond the scope of this paper to describe the operational details concerning the ways that the several methodological
variants of dietary intake surveys are conducted.     These are
available in the literature (Marr 1971).     Attention will be
focused on the planning and organization that is supposed to
precede the operative stage.

In the lower part of figure 1 are indicated, enclosed in
boxes, the several stipulations that need to be set in advance.
These are essential for guiding the choice of the appropriate
technique and for ensuring that meaningful results are obtained.
Results will be called meaningful in this context if they allow:
(a) proper evaluation of the intakes of the individual or group,
i.e., when the measured intake accurately and reliably represents the true intake; (b) comparison of one individual with
another, i.e., when subjects can be consistently characterized
in terms of their energy intakes, for example, classified in any
successive quantile, with a given probability of misclassification;
(c) identification of the role of the diet in relation to selected
criteria or characteristics, i.e., when we can establish the
existence of correlations between energy intake and, say,
obesity, or energy expenditure, or body weight.

The lack of clear and preliminary definition of the indicated
stipulations might cause a loss in the discriminatory power of
the research design.     The consequences of a low discriminatory
power are manyfold;     for example, it will interfere with the
ability to establish, in cases of absence of strong associations,
whether we are dealing with true negatives or with the
consequences of a faulty or inadequate design.     The discriminatory power is determined by the size of the intra-
individual and by the ratio of intra- to inter-individual variability.
A high intra-individual variation or a high ratio raises the risk
of misclassification of the subjects to unacceptably high levels,
and obscures the investigated relationships by reducing the
accuracy of the estimate of correlation coefficients.

One first source of variability may be identified in the
methodological error associated with the instrument.     This is,
under optimum conditions, a fixed characteristic of the method
and should be estimated and known.     It affects the accuracy
of the results.

In the case of food-intake surveys, the measurement error

is the compounded expression of a number of possible errors
deriving from a variety of sources such as variation in the
true or apparent content of energy and nutrients in foods
(Eagles et al. 1966, Kinsella et al. 1975), variation in the
digestibility factor affecting the metabolic availability of
absorbed energy (Calloway & Kretsch 1978), inability to perfectly replicate the characteristics of the ingested foods in the
process of sampling dietary duplicates for chemical analysis
(Chan Chim Yuk et al. 1975), out-datedness or non-perfect
adherence of tables of food composition, and true modifications
in the eating behaviour of the subjects as a consequence of
being observed.

Although it is impossible, by definition, to achieve absolute
accuracy of field data collected on free-living populations,
great care should be nevertheless directed to minimize all
known sources of measurement errors.   An important point is
that a high degree of accuracy of the findings does not
guarantee an accordingly high level of reliability. Systematic
errors may be present which, while not affecting the accuracy,
may systematically bias the findings.

The other major components of variability are the true
biological intra- and inter-individual variations which represent
an important source of trouble, directly affecting the reliability
of the assessment.

Intra- and inter-individual variability of energy intakes are
not of fixed character and may differ between individuals and
groups.   The variability observed in the above cited dietary
studies (table 1) is illustrated in table 2, expressed as
coefficients of variation ($CV_i$, intra-individual; $CV_g$, inter-
individual) and as their ratio.   These values are comparable
with data reported in the literature (Young et al. 1953),
Hankin et al. 1967, Balogh et al. 1971, Beaton et al. 1979).
Both coefficients show a wide range, the lowest value being
less than half the highest one.   The highest coefficients have
been recorded in association with subsistence patterns of life,
but a strictly valid comparison is not possible because of the
different size of the groups and the different duration of the
survey (five versus seven days).   As illustrated by the
values of the ratio, the $CV_i$ is often as large and in some cases
larger than $CV_g$.   I have already pointed out that this
condition seriously affects the ability to interpret the results.
This problem may be in part overcome by increasing the number
of days of observation.   Although this practice does not
decrease the real biological intra-individual variability, it mini-
mizes the risk of misclassification of the subjects.

TABLE 2. Intra- and inter-individual coefficients of variation of energy intake.

| Group | Length of survey (d) | Coefficient of variation (%) intra | Coefficient of variation (%) inter | Ratio (intra/inter) |
|---|---|---|---|---|
| Shipyard workers | | | | |
| Italy | 7 | 17.4 | 18.8 | 0.93 |
| Farmers, middle-age | | | | |
| Italy | 7 | 14.7 | 15.4 | 0.95 |
| Finland | 7 | 16.2 | 12.6 | 1.29 |
| USA | 7 | 19.7 | 15.7 | 1.25 |
| Subsistence farmers | | | | |
| New Guinea | | | | |
| Adult males | 5 | 28.7 | 22.1 | 1.30 |
| Adult females | 5 | 32.2 | 26.2 | 1.23 |
| Children | | | | |
| Italy rural 1–5 y | 7 | 17.2 | 23.5 | 0.73 |
| New Guinea 3–5 y | 5 | 30.3 | 29.1 | 1.04 |
| New Guinea 6 –10 y | 5 | 25.6 | 25.3 | 1.01 |
| New Guinea 11–17 y | 5 | 26.9 | 28.7 | 0.94 |

However, the prolongation of a survey is not without inconveniences.    Besides cost and logistic considerations, the one
serious built-in limitation is represented by <u>acceptability</u>.    By
definition, field studies on free-living subjects are performed
on volunteers.    The question of the acceptability of the survey
is therefore far from being trivial, as it may conceivably represent a disincentive for the subjects;    the higher the degree of
co-operation required, the less acceptable the survey and thus
the less the chance of securing unbiased samples of assistance
with the population project.    In a general way, one may state
that the degree of precision of a survey is in inverse proportion to its acceptability (figure 2).

More or less sophisticated statistical support to this end is
available in the literature (Balogh <u>et al</u>. 1971, Liu <u>et al</u>. 1979,
Keys 1979).

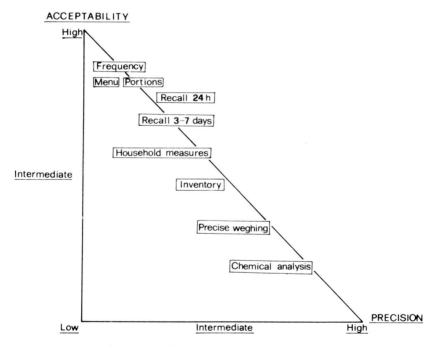

FIGURE 2.    Relationship existing between acceptability
       and precision of the various methods for assessing
       individual dietary intakes.

Applying the statistical guidelines provided by Beaton (Beaton et al. 1979) and drawing from the results of the above-mentioned surveys (table 1), two graphs have been developed off which the number of measurements required to achieve a given degree of reliability in the estimate of individual (figure 3) and of group (figure 4) values may be easily read.

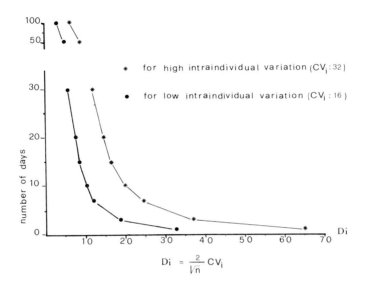

$$Di = \frac{2}{\sqrt{n}} CV_i$$

FIGURE 3.    Improvement of the reliability, at 95% confidence limits, of individual energy intake, expressed as ± percent deviation of observed mean from true mean Di, with increasing number of days of observation.    The two lines correspond to: •——• individuals with low intra-individual variability (drawn from findings on Finnish farmers; intra-individual variation coefficient $CV_i$ = 16), *——* individuals with high intra-individual variability (drawn from findings on New Guinean adult women; $CV_i$ = 32).

The reliability is here expressed as per cent deviation of the observed mean from the <u>true</u> mean at 95% confidence limit. The highest and the lowest intra-individual coefficients of variation recorded in the mentioned studies, those displayed respectively by New Guinean women and Finnish farmers, have been used in the calculations.

Figure 3 indicates that, for the individual, surveys of short duration are associated with a very large deviation of the findings from the true value. With one-day surveys we may expect 32% and 64% deviation for individuals characterized respectively by a low and a high variability of their day-by-day eating behaviour, corresponding to about ± 1000, or ± 2000 kcal, for an average intake of 3000 kcal. Such results imply

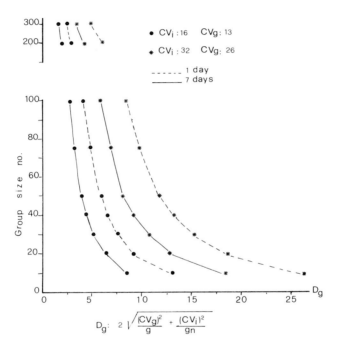

FIGURE 4. Improvement of the reliability, at 95% confidence limits, of observed group mean energy intake, expressed as ± percent deviation Dg from true mean with increasing group size and number of days of observation. The lines correspond to: ●, a group with low intra-individual variability (drawn from our findings on Finnish farmers; $CV_i$ = 16; $CV_g$ = 13; * a group with high inter-individual variability (drawn from findings on New Guinean adult women; $CV_i$ = 32; $CV_g$ = 26). The broken lines refer to one day surveys and the continuous lines to seven days surveys.

an unacceptably high risk of misclassification of the individual, even under the best conditions of rather low biological variation. Prolonging the observation to seven days improves significantly the deviation from the true mean, but it is fairly obvious that, thereafter, the improvement, i.e., the reduction of the deviation from the true mean, proceeds at a very slow pace. Increasing the observation period to 15 days brings the deviation down only from 12% to 8% (or from 24% to 17%).

Alternatively, the interest of the study may be in discriminating between groups rather than between individuals. Figure 4 has been developed for this purpse and illustrates how, by manipulating either the group size or the number of days of observation, or both, it is possible to improve the reliability of the results. The same groups with the highest and the lowest inter-individual coefficient of variation have again been used. The equation in the lower part of the graph, derived from Beaton et al. (1979), shows that the accuracy of the results is affected by the $CV_i$ and $CV_g$, as well as by the group size (g). The graph indicates that a similar degree of accuracy may be achieved either by increasing the number of observations (n) or the size of the group. For example, to stay within ± 10% deviation from the true group mean, we may elect to study 20 people for seven days or 40 people for one day. The choice will depend upon considerations relative to the acceptability, cost and logistics of longer surveys versus larger samples. Based exclusively on economic considerations, it has been calculated that, for groups with an estimated $CV_i/CV_g$ ratio of about one, a one day survey on a larger sample would be preferable to a longer survey on a smaller group if the cost of the survey is at least four times that of recruiting new subjects. Conversely, the choice of a longer survey on less subjects would be advantageous in the case where the $CV_i/CV_g$ ratio is as high as five and the cost of recruiting is less than one-fourth that of repeating the survey for more days (Beaton et al. 1979).

Whichever has been the choice, one-day or several-day surveys, the next pertinent question is: which days? The observed variability of the individual energy intake is consistent with the notion that people eat differently on different days, and it would not be surprising that a more or less pronounced weekly cycle existed.

There is not much to be found in the literature on this question, and the meagre experimental evidence is rather conflicting. Beaton has found a significant day-of-the-week effect, with increased intakes on Sundays among women, but not among men (Beaton et al. 1979). Chalmers et al. (1952) report that out of 15 groups of men, women and students of various grades, only college students showed an infra-week variation of intake, eating less at weekends than during the week. An analysis of our own material has yielded similarly inconclusive results. A marked decrease in the energy intakes during weekends (no-work days) was found among Finnish farmers, some 300 kcal, or approximately minus 10% (table 3). American farmers exhibited only a modest decline in their weekend energy intake, minus 100 kcal, while there was very little infra-week difference in the diets of Italian farmers. Contrasting behaviour was observed in the group of Italian shipyard workers who displayed a modest but significant increase of their energy intakes (150 kcals = +5%; p = 0.006) at weekends (table 4).

TABLE 3.   Mean daily intake of total energy, and % energy from alcohol, of three groups of farmers on working and non-working days.

|  | Total Energy (kcal) | | Alcohol Energy (% of total) | |
|---|---|---|---|---|
|  | Work days | No-work days | Work days | No-work days |
| Italian farmers | 3718 | 3675 | 14 | 15 |
| Finnish farmers | 3354 | 2962 | 1 | 1 |
| US farmers | 3316 | 3211 | 1 | 0.5 |

These results suggest that some subjects may indeed have a distinct weekly trend in the eating pattern, but that it is not self-evident who does and who does not. This has important practical implications, as it indicates that the conceivable presence of a day-of-the-week effect needs to be investigated in all surveys in order to correctly account for the right proportion of weekdays to be represented in the survey.

3.   INTERPRETATION OF RESULTS

    Once the energy-intake data have been collected, supposedly with the required accuracy and reliability, the next step is to understand what they mean. One of the most widely accepted and current procedures, especially if the study unit is the

TABLE 4. Mean energy intake of 150 Italian shipyard workers on work/no-work days.

|  | Work days (kcal) | No-work days (kcal) | Difference % |
|---|---|---|---|
| Energy intake (Mean ± S.E.M.) | 3198 ± 61 | 3354 ± 82 | + 5 |

Analysis of Variance

| Components of variance | df | Sum of squares | Mean squares | F ratio | Significance of F |
|---|---|---|---|---|---|
| Work/no-work days | 1 | 5 245 592 | 5 245 592 | 7.556 | 0.006 |
| Residual | 1045 | 725 440 492 | 694 201 |  |  |
| Total | 1046 | 730 686 080 |  |  |  |

group and the purpose is to reach an estimate of the energy
state of the community, consists in comparing the collected data
with nutritional standards of reference, to assess how far above
or below them the subjects fall.    Age and sex-specific energy
requirements of reference individuals are available (WHO 1973),
requiring a couple of stepwise adjustments to fit the sample
population.    Several pitfalls, however, appear to be associated
with this practice as well as a certain degree of circular
reasoning.
    Most of the readers will probably be familiar with the
manipulations needed to match factorially the study subjects
with the theoretical standards, which are expressed in terms
of the reference man.    These manipulations are intended to
account for a number of factors which are considered to be
implicated in determining the level of energy expenditure and,
thus, likely to influence energy intake.    The factors which are
usually taken into account, for normal healthy  subjects, are
age, sex, body mass and physical activity.
    No major problem is apparently associated with sex; there
is no doubt that men and women differ markedly in their energy
requirements, although this phenomenon has not yet found a
satisfactory explanation (Durnin 1969).    For example, it is not
fully understood why this difference is greatly reduced but
still not totally suppressed when lean body mass (LBM) is used
as the term of reference instead of weight (Durnin 1976).    However,
the sex specificity of energy requirements may be momentarily
taken for granted and results interpreted accordingly.
    The next adjustment factor is age.    I shall here consider
only the problems associated with ageing, ignoring the even
more complex aspects of energy needs during the developmental
stages of life.
    A decrease in energy requirement is considered to occur
physiologically with advancing age;   WHO/FAO standards
contemplate a 5% decrease for each decade between the ages of
40 and 59, and 10% thereafter (WHO 1973).    The logic of this
reduction rests upon an assumed decline in physical activity
(a point which is discussed later) and in part on a decrease
in the metabolically active compartment of the body, reflected
in a fall in the basal metabolic rate.
    An analysis of our own material, looking for evidence of
such a decline, reveals that this may not be necessarily and
invariably true.    No correlation was found between energy
intake and age in three out of four groups of New Guineans
aged between 18 and 60 years (Norgan et al. 1974).    A
similar absence of the ageing effect on energy intake in 150
shipyard workers aged 22 to 55 years is shown in figure 5.
Of course, it remains to be verified how representative are

these groups from the point of view of their relation between ageing processes and energy metabolism. It is legitimate, however, to conclude that, at least in these cases, the advocated adjustment of the energy requirement on the basis of advancing age appears to be unjustified and would not have ensured a better adherence to the real-life situation. The universality as well as the appropriate timing of the age correction appears, thus, to be questionable. Other authors have expressed a similar position, remarking that "...the consensus of opinion (on the influence of ageing on energy expenditure and thus, in turn, on energy intake) is formed from the most meagre experimental evidence" (Durnin 1969).

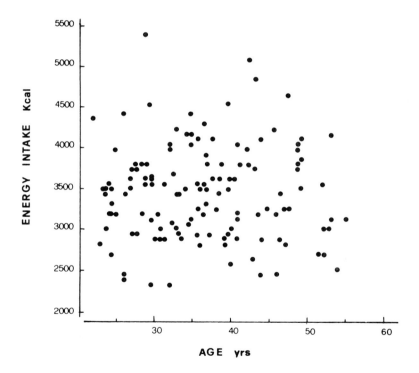

FIGURE 5. Relation of mean daily energy intake and age for 150 Italian shipyard workers, indicating absence of an ageing effect on food intake in the observed age span.

*Energy and Effort*

The next variable for the adjustment of energy requirement is <u>body weight</u>, the rationale being that it occupies a prominent position in affecting energy expenditure, and thus intake.    In this case also, the evidence is far from being unequivocal.

Figure 6 shows the results of our study on shipyard workers. Mean seven-day energy intakes have been plotted against gross body weight.    The scatter of the data appears to be very wide although a weak but significant correlation (r = 0.32) is present. A somewhat better correlation with much less scatter (r = 0.45) is obtained when the data are plotted against LBM, indicating that the latter is a slightly better somatic standard of reference (figure 7).    Still, even in this case, less than one-fifth of the energy intake appears to be accounted for by the metabolically active compartment of the body.

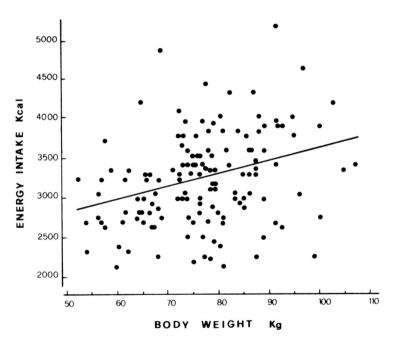

FIGURE 6.    Relation of mean daily energy intake on body
weight for a group of Italian shipyard workers (n = 150;
y = 2067 + 15.37 x; r = 0.323; p = 0.004).

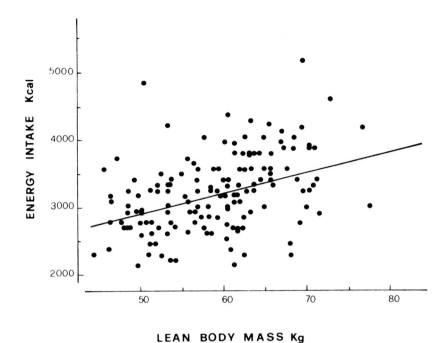

**LEAN BODY MASS Kg**

FIGURE 7.    Relation between energy intake and lean body
mass, for a group of Italian shipyard workers. (n = 149;
y = 1118 + 36.02 x; r = 0.453; p = 0.001).

Similarly low and somewhat erratic correlation coefficients
have been obtained on the other groups of subjects, ranging
from r = 0.45 for American farmers to r = 0.63 for the Italian
farmers and r = 0.04 for the New Guinean women (table 5).
When intake was correlated with LBM, instead with body weight,
the correlation was only marginally improved, and only in some
cases.
Similar findings have been reported by several other authors
(Durnin 1965, Durnin 1972, Durnin 1973, Edholm 1973).  A
satisfactory interpretation of this unexpected lack of correlation
is not available, and the whole issue remains confused and
controversial (Durnin 1969, Durnin 1972).

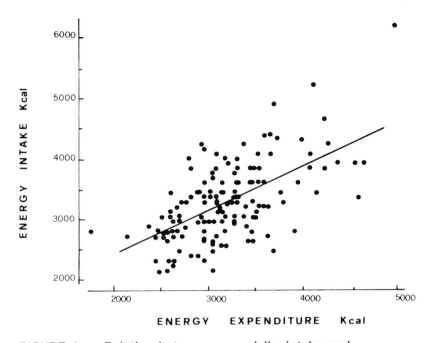

FIGURE 8.    Relation between mean daily intake and
       expenditure of energy for a group of shipyard workers
       (n = 149; y = 916 + 0.738 x; r = 0.610; p = 0.001).

    The last variable to be taken into consideration is physical
activity.    It occupies the first or second most important place
as a single determinant of energy expenditure.    It is thus
legitimate to expect that it occupies a prominent position in
relation to energy intake.    In practice, in this case also the
correlations between energy intake and expenditure turn out
to be rather poor, casting some doubt on our ability to understand the dynamics of energy metabolism in man.    First of
all, direct assesment of physical activity and energy output
of free-living individuals is neither easily nor frequently
performed.    Most dietary surveys provide, at the most, an
approximate indication of the job title and only occasionally a
crude attempt to describe the time allocation.    Job titles,
however, have been found to be often poor indicators of
activity, and the reliability of the assessment of the level of
energy expenditure that may be obtained by such indirect means

TABLE 5.    Correlation coefficients (r) between mean daily
energy intake (kcal) and body mass (kg).

| Group | No. | Correlation coefficient r |
|---|---|---|
| Italian farmers | 24 | 0.63 |
| Finnish farmers | 29 | 0.29 |
| American farmers | 21 | 0.45 |
| New Guinea coastal farmers | | |
| Men | 51 | 0.43 |
| Women | 69 | 0.04 |
| New Guinea Highland farmers | | |
| Men | 43 | 0.16 |
| Women | 41 | 0.16 |

has been questioned (Norgan & Ferro-Luzzi 1978).    We are thus
confronted with the inescapable fact that any correction for
the variable physical activity performed on this basis is likely
to have an unacceptably wide margin of unreliability and to
involve often a certain amount of circular reasoning.
    The other major problem in this context is represented by
the fact that, even when physical activity and energy output
are carefully and directly assessed, concurrently with the
measure of energy intake, the agreement between intake and
expenditure tends to be good only when average values
obtained for prolonged periods are compared (Edholm et al.
1970, Edholm 1977).    When the same results are analysed on
the individual level, the agreement is poorer, and drops even
further when the individual day-by-day values are analysed
(Edholm 1977).
    Those of our own data which lend themselves to this
analysis show that group averages may in fact agree more or
less closely, and that negative balances as low as minus 401
kcal may occur under apparently normal conditions (table 6).
The highest correlation coefficient, r = 0.61, recorded in the

TABLE 6. Comparison between mean daily intake and expenditure of energy and correlation coefficients.

| Group | No. | Energy intake (kcal) | Energy expenditure (kcal) | Difference (kcal) | Correlation coefficients r |
|---|---|---|---|---|---|
| Shipyard workers | 149 | 3242 ± 50 | 3147 ± 40 | + 95 | 0.61 |
| New Guinea coastal | | | | | |
| males | 42 | 2006 ± 73 | 2347 ± 60 | - 341 | 0.17 |
| females | 40 | 1429 ± 72 | 1830 ± 41 | - 401 | 0.17 |
| New Guinea Highland | | | | | |
| males | 40 | 2532 ± 77 | 2573 ± 42 | - 41 | 0.00 |
| females | 38 | 2099 ± 74 | 2247 ± 38 | - 148 | 0.05 |

group of shipyard workers, indicates that energy intake accounts, at its best for only up to one-fourth of total energy expenditure (figure 8). The correlation between individual intake and expenditure becomes very tenuous ($r = 0.33$) when findings are analysed on a daily basis (not shown in table) confirming the results reported by other authors (Edholm 1977). These considerations lend support to the commonly held notion that it is not necessary for energy flow to balance out instantaneously. The implication of all this is that energy imbalance may exist under normal, everyday life over periods extending up to 5—7 days, and that, in the short term, energy intake may be only marginally related to energy expenditure. As a result, the basic assumption on which interpretation of food-intake surveys rests, that of a state of equilibrium between energy intake and expenditure, can be questioned.

To conclude, we are left, unfortunately, with a disappointingly long list of unresolved problems concerning our capacity to fully understand the findings of field studies of energy intake. It is hoped that these considerations do not act as a deterrent, discouraging further and much needed research, and that they prove instead to be a stimulus and a challenge to acquire the necessary expertise, to improve the methodology, and to renew the efforts to develop integrated multidisciplinary approaches.

REFERENCES

Balogh, M., Kahn, H.A. & Medalie, J.H., 1971, Random repeat 24-hour dietary recalls. American Journal of Clinical Nutrition, 24, 304-310.
Beaton, G.H., Milner, J., Corey, P., McGuire, V., Cousins, M., Stewart, E., de Ramos, M., Hewitt, D., Grambsch, P.V., Kassim, N. & Little, J.A., 1979, Sources of variance in 24-hour dietary recall data: implications for nutrition study design and interpretation. American Journal of Clinical Nutrition, 32, 2546-2559.
Calloway, D.H. & Kretsch, M.J., 1978, Protein and energy utilization in men given a rural Guatemalan diet and egg formulas with and without added oat bran. American Journal of Clinical Nutrition, 31, 1118-1126.
Chalmers, W.F., Clayton, M.M., Gates, L.O., Tucker, R.E., Wertz, A.W., Young, C.M. & Foster, W.d., 1952, The dietary record: how many and which days? Journal of the American Dietetic Association, 28, 711-717.

Chan Chim Yuk, A.W., Wheeler, E.F. & Leppington, I.M., 1975, Variations in the apparent nutrient content of foods: a study of sampling error. British Journal of Nutrition, 34, 391-396.

Durnin, J.V.G.A., 1965, Somatic standards of reference. In Symposia of the Society for the Study of Human Biology, 6, Human Body Composition: Approaches and Applications (Pergamon Press), p. 73.

Durnin, J.V.G.A., 1969, Energy expenditure in relation to age, sex, body weight, and physical activity. In Excerpta Medica, International Congress Series no. 213, Proceedings of the Eighth International Congress on Nutrition, p. 321.

Durnin, J.V.G.A., 1972, Unresolved problems in establishing international standards for energy requirements. In Nutrition, 1, edited by A. Chavez, Bourges, H. and S. Basta (Basel: Karger), p. 35.

Durnin, J.V.G.A., 1973, Body weight, body fat and the activity factor in energy balance. In Régulation de l'equilibre énergé- tique chez l'homme, edited by M. Apfelbaum (Paris: Masson) p. 141.

Durnin, J.V.G.A., 1976, Sex differences in energy intake and expenditure. Proceedings of the Nutrition Society, 35, 145-154.

Durnin, J.V.G.A. and Ferro-Luzzi, A., 1982, Conducting and reporting studies on human energy intake and output: sug- gested standards. American Journal Clinical Nutrition (in the press).

Eagles, J.A., Grant Whiting, M. and Olson, R.E., 1966, Dietary appraisal. Problems in processing dietary data. American Journal of Clinical Nutrition, 19, 1-9.

Edholm, O.G., 1973, Energy expenditure and food intake. in Régulation de l'Equilibre Energétique chez l'homme, edited by M. Apfelbaum (Paris, Masson et Cie), p. 51.

Edholm, O.G., 1977, Energy balance in man. Journal of Human Nutrition, 31, 413-431.

Edholm, O.G., Adam, J.M., Healy, M.J.R., Wolff, H.S., Goldsmith, R. and Best, T.W., 1970. Food intake and energy expenditure of army recruits. British Journal of Nutrition, 24, 1091-1107.

Ferro-Luzzi, A., Norgan, N.G. and Durnin, J.V.G.A., 1975, Food intake, its relationship to body weight and age, and its apparent nutritional adequacy in New Guinean children. American Journal of Clinical Nutrition, 28, 1443-1453.

Ferro-Luzzi, A., Iacono, J.M. and Dougherty, R.A., (in pre- paration), Pilot epidemiological study of thrombosis: The diets of middle-age farmers in Italy, Finland and USA.

Ferro-Luzzi, A., Ferrini, A.M., Colloridi, P. and Sette, S. (unpublished data).

Garn, S.M., Larkin, F.A. and Cole, P.E., 1978, The real

problem with 1-day diet records. American Journal of Clinical
Nutrition, 31, 1114-1116.
Hankin, J.H. Reynolds, W.E. and Margen, S., 1967, A short
dietary method for epidemiologic studies. II Variability of
measured nutrient intakes. American Journal of Clinical
Nutrition, 20, 935-945.
Keys, A., 1979. Dietary survey methods. In Nutrition, Lipids,
and Coronary Heart Disease, edited by R. Levy, B. Rifkind,
B. Dennis, and N. Ernest. (New York, Raven Press), p. 1.
Kinsella, J.E., Posati, L., Weihrauch, J. and Anderson, B.,
1975, Lipids in Foods: problem and procedures in collating
data. Critical Reviews in Food Technology, 5, 299-324.
Lechtig, A., Yarbrough, C., Martorell, R., Delgado, H. and
Klein, R.E., 1976, The one-day recall dietary survey: a
review of its usefulness to estimate protein and calorie intake.
Archivos LatinoAmericanos de Nutricion, 26, 243-271.
Liu, K., Stamler, J., Dyer, A., McKeever, J. and McKeever, P.,
1978, Statistical methods to assess and minimize the role of
intra-individual variability in obscuring the relationship
between dietary lipids and serum cholesterol. Journal of
Chronic Disease, 31, 399-418.
Marr, J.W., 1971, Individual dietary surveys: purposes and
methods. World Review of Nutrition and Dietetics, 13, 105-164.
Norgan, N.G., Ferro-Luzzi, A. and Durnin, J.V.G.A., 1974,
The energy and nutrient intake and the energy expenditure
of 204 New Guinean adults. Philosophical Transactions of the
Royal Society of London, 268, 309-348.
Norgan, N.G., and Ferro-Luzzi, A., 1978, Nutrition, physical
activity, and physical fitness in contrasting environments. In
Nutrition, Physical Fitness, and Health, edited by J. Parizkova,
V.A. Rogozkin (Baltimore: University Park Press), p. 167.
WHO, 1973, Energy and Protein requirements. Report of a Joint
FAO/WHO ad hoc expert committee. World Health Organization
technical Report Series no. 522, Geneva.
Young, C.M., Franklin, R.E., Foster, W.D. and Steele, B.F.,
1953, Weekly variation in nutrient intake of young adults.
Journal of the American Dietetic Association, 20, 459-464.

# HUMAN ENERGY STORES

N.G. Norgan

Department of Human Sciences,
University of Technology,
Loughborough,
Leicestershire LE11 3TU

## 1. INTRODUCTION

Information on energy stores is useful in the assessment of nutritional status, the identification of the limiting factors in physical work capacity and the calculation of biomass data and the energy flow in human systems.

## 2. DESCRIPTION OF ENERGY STORES

The amount and location of the energy reserves of a young adult male are shown in table 1.

TABLE 1. The energy reserves of a young adult male.

| | Whole body reserve | | Adipose tissue reserve | | Skeletal muscle reserve | |
|---|---|---|---|---|---|---|
| | (kg) | (MJ) | (kg) | (MJ) | (kg) | (MJ) |
| Triglycerides | 10.5 | 390 | 10.0 | 370 | 0.5 | 20 |
| Protein | 5.0 | 85 | 0.5 | 9 | 4.0 | 70 |
| Carbohydrate | 0.5 | 9 | – | – | 0.3 | 5 |
| Total | | 484 | | 379 | | 95 |

Energy reserves consist principally of fat, which contributes 80% of the total. Mobilizable protein contributes about 18% and carbohydrates the remainder. Most of the fat (98%) is deposited

139

in adipose tissue and the mobilizable protein is found mainly in skeletal muscle, which also holds two-thirds of the carbohydrate reserves. Small reserves of carbohydrate and protein also occur in the liver. Hence, fat is the most important form of energy reserve.

Passmore (1965) has suggested that "A reserve may be defined as material available for use in an emergency but whose use is attended with some impairment of health and physiological function", but loss of a store causes no such impairment. This distinction is used in the following discussion of body fat content. Fat is a store because its loss causes no impairment.

### 3.    ASSESSMENT OF ENERGY STORES

The techniques for assessing energy stores in vivo are (a) measurements of body composition, (b) skinfold thicknesses and (c) body weight. They all depend upon assumptions which give rise to an unknown degree of uncertainty.

### 3.1   Body composition

Body composition measurements allow a description of the body as either fat plus fat-free mass (FM + FFM) or adipose tissue plus lean body mass (AT + LBM). There is considerable confusion in the use and understanding of these terms. Fat plus fat-free mass can be considered a chemical description of the body, while adipose tissue and lean body mass constitute an anatomical model. According to this definition, the two should not be combined. It may be erroneous, it is certainly misleading, to describe the body as fat plus lean body mass.

### 3.1.1 Methods of measuring body composition

Densitometry. The FM and FFM can be calculated from the body density (Db) = body mass/body volume. Body volume is obtained by water displacement. An allowance must be made for the volume of air in the lungs which is best measured at the time of water displacement. The proportion of the body weight as fat, F, can be calculated from the Db provided the densities of FM and FFM are known and are constant.

$$\frac{1}{Db} = \frac{F}{Df} + \frac{1-F}{Dffm}$$

Df is taken as 900 kg/m³ and Dffm as 1100 kg/m³ and F = (4950/Db) − 4.5. Fat mass = F × body mass.

The values for Df and Dffm were derived by Siri (1956) and there is considerable evidence to support them. The uncertainty of the FM estimate due to biological uncertainty and technical error is ±4% of the body weight as fat (Siri 1961).

Whole-body potassium monitoring. Potassium is situated in FFM, mainly in cells. From the average content of potassium in FFM and the total potassium, the amount of FFM and hence FM can be calculated. The average potassium content of FFM is taken to be 68.1 mmol/kg in males (Forbes & Hursh 1963) and 65.2 females (Forbes et al. 1968). There is evidence that this may change with muscular development, obesity and age (Womersley et al. 1976).

Whole-body potassium content can be determined with an accuracy of about 4% by counting the gamma radiation emitted by the naturally-occurring long-lived isotope [40]K in the body at 0.012% of total potassium.

Measurement of total body water. Fat is anhydrous and body water is located in FFM. If the proportion of water in FFM is known it is possible to calculate the amount of FFM from a measurement of total body water (TBW). There is a variety of techniques for measuring TBW (Sheng & Huggins 1979), most of which are based on the dilution principle. The validity of the determination of FFM and hence FM depends not only on the accuracy of the measurement of TBW but also on the variability of the water content of FFM, usually equal to 73 ± 3% (mean ± S.D.).

Several studies have compared the results obtained from the simultaneous measurement of body composition by these three approaches. In general, the agreement is good.

Each method assumes a constant composition of FFM. The tissue gained or lost when energy stores change consists primarily of fat but also includes water, protein and potassium which cause the FM composition to change. How important is this when measuring changing stores? It can be shown that body weight needs to change by 35% during weight gain and 15-25% during weight loss before the change in density of the fat-free mass exceeds the uncertainty of the value taken. The calculation of body fat from density is relatively insensitive to changes in hydration compared to that from total body water (Behnke 1961), and Grande (1968) has convincingly demonstrated the erroneous results that may be obtained with TBW and [40]K measurements during periods of high energy deficits. Thus, densitometry is

less affected and no significant error is introduced by the degree
of weight change usually found in populations.

### 3.2.  Skinfold thicknesses

Measurements of skinfold thickness (skin plus subcutaneous fat)
are a simple method of assessing body fat.   The measurements
can be standardized and are reliable in the hands of trained investigators.   There is extensive normative data on skinfolds at
all stages of growth and development (Tanner & Whitehouse 1975,
National Center for Health Statistics 1977).   It is generally believed that the proportion of fat situated subcutaneously is about
half of the total body fat content.   This is not well established
(Durnin & Womersley 1974, Brown & Jones 1977) and variations
in the skinfold thickness at one or more sites may not reflect
changes in energy stores.   Skinfold thicknesses are negatively
correlated wth body density and a considerable number of equations to estimate body density and hence fat mass have been
derived.

In principle, these equations are only valid for the populations
on which they are drawn up.   There may be differences in the
density—skinfold relationships, owing to varying proportions of
fat situated subcutaneously or compressibility of skinfolds in
different populations, and as a result of age, sex or plane of
nutrition (Durnin & Womersley 1974).   In practice, in terms of
fat content, the errors involved are not large, even though they
may be systematic and reach levels of statistical significance.

The use of estimation equations drawn up on European subjects has been found to be appropriate for Indians (Sen &
Banerjee 1958), three out of four groups of Indian men (Jones
et al. 1976), Chilean men (Apud & Jones 1977) and Taiwanese
(Parizkova 1977) but not Eskimos (Shepherd et al. 1973) and
possibly Indian women (Satwanti et al. 1977). We have found
that Durnin & Womersley's equations estimated % fat accurately
in New Guinean highland and coastal women and in young men,
except in coastal subjects where the difference was 2%, but in
older men the equations were inappropriate, giving a mean percentage fat of 14% compared to the measured mean of 10%
(p<0.001) (Norgan, Ferro-Luzzi & Durnin, unpublished).

It is surprising that there is agreement between measured and
predicted fat contents, as many of the populations are leaner
than Europeans and fall at the lower end of their range of skinfolds.   Although more information is required, the use of skinfold thicknesses to calculate fat contents using prediction equations drawn up on Europeans appears to be justifiable.

## 3.3 Body weight

The most variable component of the body weight is the fat content, and the simplicity of the measurement of weight ensures that it is a common measure of energy stores. However, weight is not a reliable indicator of fat content. This was first demonstrated by Welham & Behnke (1942) in American football players who would have been classified as overweight or obese on the basis of their weight for height but who had very low fat contents. Their extra weight was due to muscle. Also, it is possible to have a normal-weight, fat person, as many sedentary individuals may well be. Similarly, weight changes are a poor indication of changing energy stores. During weight gain or loss the composition of the tissue gained or lost varies, particularly in the first 3—4 weeks, and it is difficult to convert weight changes to energy changes in this period. Indeed, weight may be constant or increasing even when energy stores are being depleted.

## 4. VARIATIONS IN ENERGY STORES

Energy stores vary in different individuals and throughout life in any individual. Variations arise from an imbalance between intake and expenditure of energy which may occur over a range of intakes or expenditures. However, various circumstances usually have a predictable outcome; e.g., high levels of intake, low levels of expenditure, aging and acculturation are usually associated with increasing energy stores.

## 4.1 Intake of energy

There have been a number of well controlled studies on the effects of undernutrition on energy stores in individuals of normal weight (Keys et al. 1950) and in the obese (Garrow 1974 but fewer studies on the effects of voluntary over-feeding (Keys et al. 1955, Goldman et al. 1975, Norgan & Durnin 1980). The rate of change of energy stores in the general population is much slower than those induced in the laboratory or obesity clinic. Fat-mass gain in males is about 3—4 kg per decade between the ages of 25—65 years. This is equivalent to a gain of 30—40 kJ/day. Even in the massive obesity developed by body-weight record holders, the rate would only be about 600—800 kJ/day. Supplies of food may be sharply reduced through natural or man-made disasters, but usually the onset of famine is gradual.

Therefore, the adjustments to changes in energy intake in the population may differ from those in the laboratory.

## 4.2.  Expenditure of energy

If an engineer were to design the human body he would probably increase the energy stores in those with high levels of energy expenditure.   In nature, the reverse obtains.   Athletes have low proportions of body fat (Parizkova 1977), particularly those in endurance events.   Indeed, increasing the energy stores decreases the level of habitual physical activity (Norgan & Durnin 1980).

Prescriptions for developing and maintaining fitness in healthy adults by rhythmic aerobic exercise such as jogging and running lead to changes in energy stores (American College of Sports Medicine 1978).   Body weight and FM are generally reduced, while FFM remains constant or increases slightly provided an expenditure of 0.8—1.3 MJ (200—300 kcal) per session occurs.

Large energy stores are common in individuals of affluent groups and populations.   Mayer and co-workers, after surveying the body weights and energy intakes of groups of varying levels of habitual activity, suggested that the imbalance arises because activity levels are too low (Mayer et al. 1956).   This attractively simple suggestion is often quoted in textbooks.   However, Garrow (1978) has pointed out a bias in the interpretation of the data in man, and the results in rats (Mayer et al. 1954) appear to require explanation, as in the more active rats the food consumption is increased more than the cost of the extra activity. In addition, there appear to be sex differences in the response to exercise of rats.   Food intake is increased in proportion to extra energy expenditure in females, as Mayer found, but not in males (Oscai 1973).

In man, a considerable proportion (50%) of the total daily energy expenditure results from resting metabolism, and the range of habitual physical activities is associated with differences in daily energy expenditures of only 50%; e.g., the mean daily energy expenditure of elderly retired men is 9.7 MJ (2330 kcal) per day and that of coal miners is 15.3 MJ (3660 kcal) per day (Durnin & Passmore 1967).   Perception of differences in activity is much greater than the differences in energy expenditure.   This seems to have resulted in an over-emphasis of the effects of physical activity in energetic terms (Norgan & Ferro-Luzzi 1978).

4.3.   Age and sex

    There are difficulties in the quantitative description of energy
stores before puberty.   There are few normative data and different techniques have been used by different authors.
    The 50th centiles of triceps and subscapular skinfolds of boys
and girls decrease between 1—8 years (Tanner & Whitehouse
1975).   In girls, fat is deposited at both sites before, during
and after adolescence.   This also holds for the subscapular site
in boys but at the triceps site a decrease occurs during adolescence.   At all ages, boys are leaner than girls, a pattern that
continues throughout life.   The results of the Ten State Nutrition Survey on 15 000 white, lower-income subjects suggest a
pre-school decrease in fatness and a noticeable later adolescent
loss in boys (Garn & Clark 1976).
    Two of the more extensive series of FM data from body compo-
sition measurements are those of Burmeister & Bingert (1967) and
Forbes & Amirhakimi (1970) based on $^{40}$K measurements of eight
year olds and over.   In Burmeister & Bingert's study the 50th
centiles of FM were 3 kg at 7.5 years, 6 kg at 12.5 years and
9 kg at 17.5 years in boys and 5 kg, 10 kg and 17.5 kg respectively
in girls.   Both series show a greater FM and rate of increase in
FM in girls and a mid-teens loss of FM in boys, which was not
apparent in the subjects of Novak (1963).
    Density measurements have been made in children as young as
seven years (Parizkova 1961, 1977) but FM cannot be calculated
as body weights are not given.   Cureton et al. (1975) compared
densitometric, $^{40}$K and skinfold estimates in pre-pubescent boys
and found no significant differences between mean percentage fat
estimates, but $^{40}$K estimates of fatness were lower than those
obtained by the other two methods in some individuals.   As with
adults, FFM may be estimated from anthropometric dimensions
(Lohman et al. 1975, Slaughter et al. 1978) but this does not
mean that FM can be calculated accurately from the estimated
FFM.   Longitudinal studies of FM in 96 boys between 11 and 15
years show slightly lower gains (0.3-2.8 kg depending on level
of activity) than those of cross-sectional studies (Parizkova 1968).
    Fat deposits continue to increase throughout adult life.   This
is not apparent from measurements of body weight alone.   Average
weights increase until the fifth decade and then fall (James 1976).
Weight changes are the resultant of two processes, an increasing
FM and a decreasing FFM.   The results from several series of
cross-sectional measurements of body composition in 20-65 year
olds show a marked similarity of rate of FM gain of 3—4 kg per
decade in both sexes.   This is remarkable in that different
techniques and samples have been used, densitometry by Young
et al. (1963) and Durnin & Womersley (1974), and $^{40}$K by Forbes

& Reina (1970), and as $^{40}$K may give consistently higher fat contents in older subjects (Myhre & Kessler 1966).     There are few
longitudinal date (Forbes & Reina 1970) and the measurements of
older subjects are probably influenced by survival of leaner individuals and subject selection.

The age changes may be more complex than this description
suggests, because of the problem of interpreting the results of
the different methods and because the rates may differ in middle
age and old age.    The changes in skinfolds over this age range
are twice as great in women as in men (Garn & Clarke 1976, Shephard 1978) but FM changes are similar.    This suggests that the
proportion of fat situated subcutaneously increases more in
women than in men, which is contrary to what is expected (Durnin & Womersley 1974, Shephard 1978), or that the composition
of FFM is changing more in women.

For the purposes of describing energy stores, FM is clearly a
better indicator than body weight.    A 3 kg gain in FM represents
the storage of 110 MJ ($27 \times 10^3$ kcal).    The fall in FFM of 1 kg
per decade represents a small loss of energy of 4 MJ ($1 \times 10^3$
kcal).    Energy stores double between 25—65 years.    Ideally,
body weights should not increase with age because of the association of increased morbidity and mortality with overweight.    The
concern should probably be with fat content rather than weight.
If this is so, to maintain a constant fat content between 25—65
years, body weights should fall 12 kg in men and 5 kg in females.
Such changes do occur in the populations of developing countries
(Sinnett et al. 1973).

## 4.4.  Genetic factors

In some animals, defect of a single gene may cause abnormally
large energy stores, e.g. the Zucker rat and the ob/ob mouse,
but this is not known to occur in man.    Obesity – the excessive
accumulation of fat – runs in families.    The level of fatness (as
shown by skinfold thicknesses) of the child rises progressively
with the level of fatness of the parental pair (Garn & Clark 1976).
The likelihood of a child being obese is 40—50% if one parent is
obese and rises to 70—80% if both parents are obese.    Children
of obese parents are three times as fat as those of lean parents
and the fatness of siblings increases with fatness of the propositi.
However, there are many non-genetic factors common to all members of a household.    Garn et al. (1979) have found a synchrony
in long-term fat gain or loss in spouses – genetically unrelated
adults – which illustrates the cohabitational effect.    This effect
also influences dietary intakes and blood and urine metabolite

levels (Garn et al. 1979). But identical twins reared in the
same environment show less difference in body weight than do
non-identical twins and the weights of adopted children show no
relationship to their foster parents even when adopted soon after
birth.

The heritability of trunk (subscapular) and limb (triceps) skinfold thicknesses in like-sex twins 3-15 years old was determined
by Brook et al. (1975). Genetic factors appeared more important
in determining trunk fat than limb fat. Heritability was high
for both trunk and limb fat in children over ten years of age
but for younger children environmental influences appeared to
be of greater importance.

Genetic factors may also be responsible for the ethnic differences
in the site of subcutaneous fat deposition. Europeans and Euro-
Americans deposit proportionally more fat on the upper limbs than
Afro-Americans, Asiatics and Amerindians (Eveleth 1979).
Johnston et al. (1974) have suggested that racial differences in
6—17 year old Afro- and Euro-Americans in triceps thicknesses
arise mainly from hereditary factors, while those of subscapular
skinfold thicknesses mainly arise from environmental causes.
These results can be contrasted with those of Brook et al. (1975).

The increasing incidence of obesity and diabetes mellitus in
populations changing from a traditional to a Western life-style led
Neel (1962) to propose the existence of a thrifty, diabetic genotype in man that is better able to store energy, an advantage in
times of famine, but one rendered detrimental by progress.
Rothwell & Stock (1981) in a deliberately speculative review have
taken another view and described the evolutionary advantages
of leanness and variable efficiency of energy utilization. They
consider leanness to be widespread in nature but not as a result
of dietary inadequacy, as wild animals remain lean when adequate
food is available. Leanness confers greater locomotor ability
and reproductive success. The ability to vary the efficiency of
energy utilization and so produce more or less heat is beneficial
in cold acclimatization, in improving the diet and in leading to
leanness.

4.5.   Way of life

Way of life has an important influence on energy stores. Annual
changes may occur in agricultural systems that involve seasonal
variations in effort and availability of foods (Fox 1953). Urban-
ization and migration involve changes in diet and physical activity
such that overweight and obesity become more prevalent. A
striking illustration of this are the Nauruan Micronesians, who

are believed to have one of the highest per capita incomes in the world, and who are markedly obese (Ringrose & Zimmet 1979). The average energy intakes of 30 MJ in men and 20 MJ in women must be some of the highest recorded in any population group.

There has been a trend in Western countries of increasing mechanization at work and in the home, with shorter working hours, although the evidence for declining physical activity is not impressive (Edholm 1970). Whether this affects energy stores is not clear. Montegriffo (1971) compared the weights of UK adults in eight studies, four prior to 1951, and four after 1968. Men were 8 kg heavier on average but women under 50 years showed no increase. Although part of the difference may be associated with greater stature, he concluded that the male population was more obese. There is also evidence of a secular increase in body-mass index (weight/height$^2$) in Army recruits between the 1940s and 1960s (James 1976). However, Malina (1979) found little evidence to suggest large changes in body composition in adult American males since the 1940s.

Socio-economic differences in energy stores occur in most populations but in ways that preclude generalization. American children and adult men from higher socio-economic classes have larger skinfolds at virtually all ages (Garn & Clark 1976) but there is an income-related reversal in fatness in females during late adolescence. Lower socio-economic-class girls are leaner but women in this class are fatter than average. Overweight is less common in UK upper-class men than in lower classes but in FR Germany the reverse is true (James 1976), illustrating the action of cultural factors on energy stores.

4.6    Population differences in energy stores

Population differences in energy stores are the resultants of the effects of the environmental and genetic factors acting on intake and expenditure of energy. Each of these factors have been discussed but some comments on the overall effect as manifested in population differences can be made. Few measurements have been made of the body composition of populations other than American or European groups. The results for men are shown in table 2 and for women in table 3 with some typical values for Europeans.

New Guineans have lower percentage fat and FM and are more homogeneous in energy stores than Europeans. They do not exhibit the gain in stores between the third and fourth decades shown by the Europeans. The Indian men have energy stores similar to New Guineans. Some of the values calculated from

TABLE 2. The body composition of men in various populations.

| Group | n | Age (y) | Body weight (kg) | % Fat Mean | % Fat SD | Fat mass (kg) | Method and Authors |
|---|---|---|---|---|---|---|---|
| Scottish residents | 92 | 20—29 | 70 | 15 | 7 | 10 | Densitometry (Durnin & Womersley 1974) |
| | 34 | 30—39 | 80 | 23 | 5 | 18 | |
| | 34 | 40—49 | 77 | 25 | 7 | 19 | |
| New Guinean horticulturalists | 73 | 23 | 58 | 9 | 3 | 5 | Densitometry (Norgan et al. unpublished) |
| | 49 | 35 | 57 | 10 | 4 | 5 | |
| Indian soldiers | 90 | 19—35 | 57 | 9 | | 5 | Densitometry (Jones et al. 1976) |
| Indian civilians | 30 | 19—35 | 55 | 9 | | 5 | |
| Canadian Eskimos | 12 | 20—29 | 69 | 15 | | 11 | Total body water (Shephard et al. 1973) |
| | 10 | 30—39 | 71 | 12 | | 9 | |
| Columbians, rural Mild undernutrition | 11 | 33 | 52 | 18 | | 9 | Total body water (Barac-Nieto et al. 1978) |
| Intermediate undernutrition | 18 | 40 | 48 | 20 | | 8 | |
| Severe undernutrition | 18 | 38 | 43 | 15 | | 7 | |

TABLE 3. The body composition of women in various populations.

| Group | n | Age (y) | Body weight (kg) | % Fat Mean | % Fat S.D. | Fat mass (kg) | Method and Author |
|---|---|---|---|---|---|---|---|
| Scottish residents | 100 | 20—29 | 63 | 29 | 10 | 18 | Densitometry (Durnin & Womersley 1974) |
|  | 58 | 30—39 | 68 | 33 | 10 | 23 |  |
|  | 48 | 40—49 | 68 | 35 | 8 | 24 |  |
| New Guinean horticulturalists | 68 | 23 | 50 | 21 | 4 | 11 | Densitometry (Norgan et al. unpublished) |
|  | 20 | 37 | 46 | 22 | 4 | 10 |  |
| Indians – Punjabis | 65 | 18—30 | 46 | 15 |  | 7 | Densitometry (Satwanti et al. 1977) |
|  | 65 | 20—25 | 45 | 18 |  | 8 | Densitometry (Raja et al. 1977) |
| Canadian Eskimos | 12 | 20—29 | 58 | 20 |  | 12 | Total body water (Shephard et al. 1973) |
|  | 10 | 30—39 | 55 | 26 |  | 14 |  |

TBW measurements may overestimate percentage fat and FM. Schaefer (1977) has suggested that the 20-29 year old Eskimo men may have been dehydrated after hunting trips and the 15-20% fat in undernourished Columbians seems high. The Indian women have low FMs for urban subjects. The two groups were studied by different workers but at the same laboratory, and their practice of measuring lung residual volume with subjects not completely immersed may have resulted in an underestimation of fatness.

In contrast to the paucity of information on body composition, there is much data on skinfold thicknesses. The body composition of several African groups, estimated from skinfold thicknesses using the equations of Durnin & Womersley, are shown in table 4 to illustrate the effects of acculturation on body composition. There is a trend of increasing percentage of fat and FM with acculturation. The values of the participants in the First African University Games, presumably an active, well nourished educational elite group, are similar to those of urban Africans.

The results show that most adult populations in developing countries have similar percentage fats in the body: 10—12% in men and 20—25% in women. The ratios of female/male energy stores lie between 1.5—2.0 except in the older Europeans and young Eskimos.

## 5.  CHANGES IN ENERGY STORES

At first glance there appears to be little difficulty in calculating energy storage or depletion.  If intake exceeds expenditure the excess will be stored;  if expenditure exceeds intake the stores will be depleted.  In practice, the calculation is more complicated than this, because energy stores may be defended by what can be regarded as homeostatic mechanisms.

### 5.1.  Overfeeding

Firstly, energy storage is not 100% efficient as there is an energy cost of storage.  The efficiency depends on the type of tissue laid down and the substrate for deposition.  From biochemical considerations, fat deposition has an efficiency of 84% when glucose is the substrate and 97% when fat is the substrate. Protein energy deposition is 75% efficient (Davidson et al. 1979). Animal-feeding experiments indicate much lower efficiencies

TABLE 4.   The body composition of various African groups estimated from skinfold thicknesses.

| | n | Age (y) | Body weight (kg) | % Fat | Fat mass (kg) | Author |
|---|---|---|---|---|---|---|
| **Men** | | | | | | |
| Masai – tribal | 39 | all | 58 | 6 | 3 | Day et al. (1979) |
| Masai – non-tribal | 22 | all | 59 | 6 | 4 | |
| Samburu – tribal | 45 | all | 61 | 9 | 6 | |
| Samburu – non-tribal | 63 | all | 59 | 12 | 7 | |
| Tanzanian, urban | 62 | 23 | 58 | 12 | 7 | Davies et al. (1973) |
| Nigerian soldiers | 170 | 18–29 | 63 | 12 | 8 | Watson & Etta (1975) |
| Athletes of 1st African University games | 1065 | 23 | 68 | 11 | 8 | Watson & Dako (1977) |
| **Women** | | | | | | |
| Upper Volta farmers | 15 | 31 | 51 | 19 | 10 | Bleiberg et al. (1980) |
| Tanzanians, urban | 32 | 28 | 50 | 26 | 13 | Davies et al. (1973) |
| Athletes of 1st African University Games | 216 | 22 | 56 | 24 | 13 | Watson & Dako (1977) |

(Pullar & Webster 1977). Few of the calculations of changing
energy stores in man have included an allowance for the inefficiency of storage (Norgan & Durnin 1980). Secondly, body
weight increases with overfeeding. This raises the energy cost
of tasks. These two effects tend to reduce or buffer the
expected gain in energy stores.

It has been suggested that there may be a third effect tending
to reduce energy gains. This is luxus-konsumption or dietary-
induced thermogenesis (DIT), a phenomenon said to operate by
converting excess energy to heat (Miller et al. 1967). The
investigation of this phenomenon in man is complicated because
it is very difficult technically to separate this effect from the
two mentioned above and the more familiar specific dynamic action
of foods. Rothwell & Stock (1979) have suggested that DIT in
the mouse and rat may have similar underlying mechanisms to
non-shivering thermogenesis.

## 5.2. Underfeeding

As underfeeding progresses, body weight falls, reducing the
energy cost of a standard task. Volitional physical activity
decreases and there is evidence that the metabolic activity of
tissues decreases (Keys et al. 1950) perhaps by reducing protein turnover or sodium pumping which may contribute up to
25% and 50% of basal energy expenditure (Davidson et al. 1979).
In the Minnesota study of Keys et al. (1950), young men were
given half-rations, body weight and energy expenditure fell,
but after six months energy balance was re-established.

Thus, the calculation of changing energy stores from measurements of intake and expenditure is more complicated than the
measurement of static stores.

## REFERENCES

American College of Sports Medicine, 1978, Position statement on
the recommended quantity and quality of exercise for developing and maintaining fitness in healthy adults. Medicine and
Science in Exercise and Sports, 10, vii-x
Apud, E., Benavides, R. & Jones, P.R.M., 1977, Application of
physiological anthropometry to a study of Chilean male for-
estry workers. Proceedings of the International Union of
Physiological Sciences, XIII, 29.

Barac-Nieto, M., Spurr, G.B., Lotero, H. & Maksud, M.G.,1978, Body composition in chronic malnutrition. American Journal of Clinical Nutrition, 31, 23-40.

Behnke, A.R., 1961, Comment on the determination of whole body density and a resume of body composition data. In Techniques for Measuring Body Composition, edited by J. Brozek (Washington, D.C.: National Academy of Sciences), p.131.

Bleiberg, F.M., Brun, T.A., Goihman, S. & Gouba, E., 1980, Duration of activities and energy expenditure of female farmers in dry and rainy seasons in Upper-Volta. British Journal of Nutrition, 43, 71-82.

Brook, C.G.D., Huntley, R.M.C. & Slack, J., 1975, Influence of hereditary and environment in determination of skinfold thickness in children. British Medical Journal, 2, 719-721.

Brown, W.J. & Jones, P.R.M., 1977, The distribution of fat in relation to habitual activity. Annals of Human Biology, 4, 537-550.

Burmeister, W. & Bingert, A., 1967, Die quantitativen Veränderungen der menschlichen Zellmasse zwischen dem 8 und 9 Lebensjahr. Klinische Wochenschrift, 45, 409-416.

Cureton, K.J., Boileau, R.A. & Lohman, T.G., 1975, A comparison of densitometric, potassium-40 and skinfold estimates of body composition in prepubescent boys. Human Biology, 47, 321-336.

Davidson, S., Passmore, R., Brock, J.F. & Truswell, A.S., 1979, Human Nutrition and Dietetics, 7th Ed. (Edinburgh: Churchill Livingstone), p.66.

Davies, C.T.M., Mbelwa, D., Crockford, G. & Weiner, J.S., 1973, Exercise tolerance and body composition of male and female Africans aged 18-30 years. Human Biology, 45, 31-40.

Day, J., Bailey, A. & Robinson, 1979, Biological variations associated with change in life style among the pastoral and nomadic tribes of East Africa. Annals of Human Biology, 6, 29-39.

Durnin, J.V.G.A. & Passmore, R., 1967, Energy, Work and Leisure (London: Heinemann), p. 115.

Durnin, J.V.G.A. & Womersley, J., 1974, Body fat assessed from total body density and its estimation from skinfold thickness: measurements on 481 men and women aged 16 to 72 years. British Journal of Nutrition, 32, 77-97.

Edholm, O.G., 1970, The changing pattern of human activity. Ergonomics, 13, 625-643.

Eveleth, P.B. 1979, Population differences in growth: environmental and genetic factors. In Human Growth, Volume 3, Neurobiology and Nutrition, edited by F. Falkner and J.M. Tanner (London: Bailliere Tindall), p.381.

Forbes, G.B. & Amirhakimi, G.H., 1970, Skinfold thickness and body fat in children. Human Biology, 42, 401-418.

Forbes, G.B. & Hursh, J.B., 1963, Age and sex trends in lean body mass calculated from $K^{40}$ measurements: with a note on the theoretical basis for the procedure. Annals of the New York Academy of Sciences, 110, 255-263.

Forbes, G.B. & Reina, J.C., 1970, Adult lean body mass declines with age: some longitudinal observations. Metabolism, 19, 653-663.

Forbes, G.B., Schultz, F., Cafarelli, C. & Amirhakimi, G.H., 1968, Effects of body size on potassium-40 measurements in the whole body counter (Tilt-chair technique). Health Physics, 15, 435-442.

Fox, R.H., 1953, A Study of Energy Expenditure of Africans Engaged in Various Rural Activities (London University: Ph.D. Thesis), quoted in Human Biology, 2nd Ed., by G.A. Harrison, J.S. Weiner, J.M. Tanner and N.A. Barnicot (Clarendon: Oxford), p.417.

Garn, S.M.& Clark, D.C., 1976, Trends in fatness and the origins of obesity. Paediatrics, 57, 443-456.

Garn, S.M., Bailey, S.M. & Cole, P.E., 1979, Synchronous fatness changes in husbands and wives. American Journal of Clinical Nutrition, 32, 2375-2377.

Garn, S.M., Cole, P.E. & Bailey, S.M., 1979, Living together as a factor in family-line resemblances. Human Biology, 51, 565-587.

Garrow, J.S., 1974, Energy Balance and Obesity in Man (Amsterdam: North Holland Publishing Company), p. 257.

Garrow, J.S., 1978, Energy Balance and Obesity in Man, 2nd Ed. (Amsterdam: Elsevier), p.48.

Goldman, R.F., Haisman, M.F., Bynum, G., Horton, E.S. & Sims, E.A.H., 1975, Experimental Obesity in Man. In Obesity in Perspective, edited by G.A. Bray (Washington D.C.: DHEW Pub. No. (NIH) 75-708), p.179.

Grande, F., 1968, Energetics and weight reduction. American Journal of Clinical Nutrition, 21, 305-314.

James, W.P.T. (editor), 1976, Research on Obesity. D.H.S.S./ M.R.C. Group Report (London: HMSO), p.16.

Johnston, F.E., Hamill, P.V.V. & Lemeshow, S., 1974, Skinfold thicknesses in a national probability sample of U.S. males and females aged 6 through 17 years. American Journal of Physical Anthropology, 40, 321-324.

Jones, P.R.M., Bharadwaj, H., Bhatia, M.R. & Malhotra, M.S., 1976, Differences between ethnic groups in the relationship of skinfold thicknesses to body density. In Selected Topics in Environmental Biology, edited by B. Bhatia, G.S. Chhina and B. Singh (New Delhi: Interprint), pp.373-376.

Keys, A., Brozek, J., Henschel, A., Mickelson, O. & Taylor, H.L., 1950, The Biology of Human Starvation (Minneapolis: University of Minnesota Press).

Keys, A., Anderson, J.T. & Brozek, J., 1955, Weight gain from simple overeating. 1. Character of the tissue gained. Metabolism, 4, 427-432.

Lohman, T.G., Boileau, R.A. & Massey, B.H., 1975, Prediction of lean body mass in young boys from skinfold thickness and body weight. Human Biology, 47, 245-262.

Malina, R.M., 1979, Secular changes in growth, maturation and physical performance. In Exercise and Sports Sciences Reviews, Volume 6, edited by R.S. Hutton (Philadelphia: Franklin Institute Press), p.210.

Mayer, J., Marshall, N.B., Vitale, J.J., Christensen, J.H., Mashayekhi, M.B. & Stare, F.J. (1954), Exercise, food intake and body weight in normal rats and genetically obese adult mice. American Journal of Physiology, 177, 544-548.

Mayer, J., Roy, P. & Mitra, K.P., 1956, Relation between caloric intake, body weight and physical work: studies in an industrial male population in West Bengal. American Journal of Clinical Nutrition, 4, 169-175.

Miller, D.S., Mumford, P. & Stock, M.J., 1967, Gluttony 2. Thermogenesis in overeating man. American Journal of Clinical Nutrition, 20, 1223-1229.

Montegriffo, V.M.E., 1971, A survey of the incidence of obesity in the United Kingdom. Postgraduate Medical Journal, 47, suppl. 418-422.

Myhre, L.G. & Kessler, W.V., 1966, Body density and potassium 40 measurements of body composition as related to age. Journal of Applied Physiology, 21, 1251-1255.

National Center for Health Statistics, 1977, NCHS Growth curves for children, birth—18 years. United States, Department of Health, Education and Welfare, Publication Number (PHS) 78-1650, Series 11, Number 165, (Hyattsville, Md: DHEW).

Neel, J.M., 1962, Diabetes mellitus: a 'thrifty' genotype rendered detrimental by 'progress'. American Journal of Human Genetics, 14, 354-362.

Norgan, N.G. & Durnin, J.V.G.A., 1980, The effect of 6 weeks of overfeeding on the body weight, body composition and energy metabolism of young men. American Journal of Clinical Nutrition, 33, 978-988.

Norgan, N.G. & Ferro-Luzzi, A., 1978, Nutrition, physical activity and physical fitness in contrasting environments. In Nutrition, Physical Fitness and Health, edited by J. Parizkova (Baltimore: University Park Press), pp.167-193.

Novak, L.P., 1963, Age and sex differences in body density and creatinine excretion of highschool children. Annals of the

New York Academy of Sciences, 110, 545-576.
Oscai, L.B., 1973, Role of exercise in weight control. In Exercise and Sports Sciences Reviews, Volume 1, edited by J.H. Wilmore (New York: Academic Press).
Parizkova, J., 1961, Age trends in fat in normal and obese children. Journal of Applied Physiology, 16, 173-174.
Parizkova, J., 1968, Longitudinal study of the development of body composition and body build in boys of various physical activity. Human Biology, 40, 212-225.
Parizkova, J., 1977, Body fat and physical fitness (The Hague: Martinus Nijhoff), p.36.
Passmore, R., 1965, Stores in the human body. In Human Body Composition, edited by J. Brozek (Oxford: Pergamon Press), p.122.
Pullar, J.D. & Webster, A.J.F., 1977, The energy cost of fat and protein deposition in the rat. British Journal of Nutrition, 37, 355-363.
Raja, C., Singh, R. & Bharadwaj, H., 1977, Anthropometric determinations of body volume, body density and segmental volume in adult Indian women. Annals of Human Biology, 4, 565-575.
Ringrose, H. & Zimmet, P., 1979, Nutrient intakes in an urbanised Micronesian population with a high diabetes prevalence. American Journal of Clinical Nutrition, 32, 1334-1341.
Rothwell, N.J. & Stock, M.J., 1979, A role for brown adipose tissue in diet-induced thermogenesis. Nature, 281, 31-35.
Rothwell, N.J. & Stock, M.J. 1981, Thermogenesis: comparative and evolutionary considerations. In The Body Weight Regulatory System: Normal and Disturbed Mechanisms, edited by L.A. Cioffi, W.P.T. James and T.B. Van Itallie (New York: Raven Press).
Satwanti, K., Bharadwaj, H. & Singh, I.P., 1977, Relationship of body density to body measurements in young Punjabi women: applicability of body composition prediction equations developed for women of European descent. Human Biology, 49, 203-213.
Schaefer, O., 1977, Are Eskimos more or less obese than other Canadians? A comparison of skinfold thicknesses and ponderal index in Canadian Eskimos. American Journal of Clinical Nutrition, 30, 1623-1628.
Sen, R.N. & Banerjee, S., 1958, Studies on the determination of body fat in Indians. Indian Journal of Medical Research, 46, 556-560.
Sheng, H-P. & Huggins, R.A., 1979, A review of body composition studies with emphasis on total body water and fat. American Journal of Clinical Nutrition, 32, 630-647.
Shephard, R.J., 1978, Physical Activity and Aging (London: Croom Helm), p.55.

Shephard, R.J., Hatcher, J. & Rode, A., 1973, On the body composition of the Eskimo. European Journal of Applied Physiology, 32, 3-15.

Sinnett, P., Keig, G. & Craig, W., 1973, Nutrition and age-related changes in body build of adults: studies in a New Guinea highland community. Human Biology in Oceania, 2, 50-62.

Siri, W.E., 1956, Gross composition of the body. Advances in Biological and Medical Physics, 4, 239-280.

Siri, W.E., 1961, Body composition from fluid spaces and density: analysis of methods. In Techniques for Measuring Body Composition, edited by J. Brozek (Washington, D.C.,: National Academy of Sciences), p.227.

Slaughter, M.H., Lohman, T.G & Boileau, R.A., 1978, Relationship of anthropometric dimensions to lean body mass in children. Annals of Human Biology, 5, 469-482.

Tanner, J.M. & Whitehouse, R.H., 1975, Revised standards for triceps and subscapular skinfolds in British children. Archives of Diseases in Childhood, 50, 142-145.

Watson, J.D. & Dako, D.Y., 1977, Anthropometric studies on African athletes who participated in the 1st African University Games. British Journal of Nutrition, 38, 353-360.

Watson, R.S. & Etta, K.M., 1975, Anthropometric measurements and amounts of body fat and serum cholesterol in samples of three Nigerian populations. British Journal of Nutrition, 33, 141-148.

Welham, W.C. & Behnke, A.R., 1942, The specific gravity of healthy men. Journal of the American Medical Association, 118, 498-501.

Womersley, J., Durnin, J.V.G.A., Boddy, K. & Mahaffy, M., 1976, Influence of muscular development, obesity and age on the fat-free mass of adults. Journal of Applied Physiology, 41, 223-229.

Young, C.M., Blondin, J., Tensuan, R. & Fryer, J.H., 1963, Body composition studies of 'older' women, thirty to seventy years of age. Annals of the New York Academy of Science, 110, 589-607.

# CONSUMPTION-WORK DECISIONS BY
# PRODUCERS AND LABOURERS

C. Bliss[1] and N. Stern[2]

[1]Nuffield College, New Road, Oxford
[2]University of Warwick,
Coventry CV4 7AL

## 1. INTRODUCTION

This paper is concerned with preliminary and rather speculative discussion of the influence of nutritional factors on the
consumption-work decisions of producers.   It is preliminary
because the topic is one that has been neglected, at least by
economists, and it must be speculative because important evidence bearing on the matter is not apparently to hand.   Nevertheless, we hope that we may be forgiven for raising the topic,
because it is potentially one of great importance.

In a previous paper (Bliss & Stern 1978) we examined the
theoretical implications of the hypothesis that wages offered in
employment would be importantly influenced by a link between
productivity and consumption.   Our attention was then mainly
directed to employment and to wages chosen by an employer to
maximize his gain.   We mentioned the long term but were de-
terred from giving much attention to it by the observation that
most employment in the agricultural sectors of under-developed
countries is short-term in character.   In the second part of
the paper we reported a survey of the nutritional evidence as
it related to our problem, and the reader is referred there for
nutritional evidence which we shall not repeat here.

To explain our present concerns, our previous conclusions
are a useful starting point.   In answer to the question whether
it would pay an employer to pay a higher wage than the minimum
at which labour was available (so that increased consumption
might make the worker more productive) we were led to conclude
that this was unlikely to happen because:

(a)    If the work was easily monitored, the market would
establish a rate per task completed, rather than a rate per hour

worked, and no employer would have an incentive to pay more than this rate; and

(b)    If the work was not easily monitored, the benefit of paying a higher wage would mostly be of a long-term character, because typically a worker can draw on his reserves of energy to do work on one particular day and better feeding makes him more productive later but not on the same day.

For details of this argument, consult Bliss & Stern (1978).

## 2.    A FURTHER QUESTION

We can now ask, what does happen if an employer, perhaps foolishly, pays a higher wage than he needs to and gets the same work from his employee? The employee enjoys better nutrition and, considering only energy balance as we did before, we can say that the worker will gain weight. We can say this because there is for any individual an identity between energy intake and energy use such that, given basic metabolic requirements and energy expended in work, the surplus of energy absorbed (positive or negative) goes to changing body mass. This was a reason why the model based on a productivity—consumption link failed as a short-term model, because in fact there is no such strict link in the short-term – the possibility of an alteration in body weight made the link a slack one. The same cannot be said of the long-term, on which we shall focus in the present paper. If a certain individual works at a certain level and consumes at a certain level his body weight (within limits) will stabilize at a certain value. We can reasonably say, therefore, that there is a long-term constraint on the three values: C, consumption as energy intake; M, body mass; and L, work in the form of energy expended. We can express this constraint mathematically as

$$\varphi(C, M, L) \quad = \quad O \qquad (1)$$

If we now have regard to the worker or peasant as an individual making long-term decisions, he is bound by this constraint.

In traditional economic analysis the two values C and L have appeared elsewhere in the model for work-consumption choice. First, there must be some relation showing how for the individual his work is translated into command over consumption. Two

obvious ones are the wage relation for an employee and the production function for a peasant.   We take the wage-employment
case as slightly simpler, but each would serve equally well.
Ignoring saving and assuming the wage for a unit of employment
is  w,   the worker's consumption is constrained by

$$wL = C \qquad (2)$$

There is no question of  M  appearing in this equation, which
is simply an accounting identity.
   The utility function shows the worker's preferences between
work and leisure, consuming and not-consuming, etc.   Traditionally, it has been written

$$U[C,L] \qquad (3)$$

and the worker is assumed to want to maximize it.   In view of
our present argument this does not make sense.   In trying to
make  C  and  L  as pliant as possible, the worker would make
M  very small.   He must be constrained from so doing by including  M  elsewhere in the problem.   Otherwise, (1)  is
eliminated, as a suitable value of  M  ensures that it is satisfied, and the problem becomes

$$\text{Maximize} \quad U[C,L]$$
$$\text{subject to } wL = C \qquad (4)$$

The solution is characterized by

$$\frac{U_C}{U_L} = -\frac{1}{w} \qquad (5)$$

where $U_C$ and $U_L$ are the partial derivatives of  U  with respect to  C and  L.   Economically, (5) means that the worker works until the marginal disutility of work equals the value
of the wage in extra consumption.
   We remarked on the problem of light (in weight) workers and
the apparent advantages that they would seemingly enjoy in Bliss
& Stern (1978).   Naturally, this advantage does not generally
exist in reality, and what the argument ignores is that  M,
insofar as it is reflected in lean body mass, enters the production function as a variable.   Even if the worker were to be
indifferent to his weight (itself a highly doubtful assumption),
his earnings function would reflect it.   So too would the

peasant's production function.    Either way, we can now write
the earnings constraint as

$$\rho(L,C,M) = 0. \tag{6}$$

The work-consumption problem now makes sense.    The worker
cannot afford to let his weight,    M,    fall so as to push up    L
for a given value of    C    while satisfying    (1).    This would
cause his production to fall via    (6).    His problem becomes

$$\text{Maximize} \quad U[C,L]$$

$$\text{subject to} \quad \rho(L,C,M) = 0 \tag{7}$$

$$\varphi(C,M,L) = 0$$

For the solution to this problem we have

$$\frac{U_C}{U_L} = \frac{\varphi_C - \frac{\varphi_M}{\rho_M}P_C}{\varphi_L - \frac{\varphi_M}{\rho_M}\rho_L} \tag{8}$$

which may be compared to    (5)    where the right-hand side
equals    $\rho_C/\rho_L$    in this case.
     This    is already more complicated than traditional eco-
nomics and not easy to quantify, although everything on the
right hand side of    (8)    could in principle be quantified.    However, it would be more plausible to include    M    in the utility
function as well.    This view is reinforced when one notes that
M    could be any variable to do with health, and depending on
nutrition.    It only matters that it should enter the constraint
on possible    (C,L)    values for the individual.


3.    IMPLICATIONS OF THE THEORY

     As the theory is presented at a high level of abstraction,
it is not easy to see its implications or assess its importance.
However, some are not difficult to discern.    Perhaps the most
important is the suggestion that in given production conditions
and given preferences (no doubt much influenced by culture)
there is for any community an optimum level of nutrition and a
consequent long-term average weight, which when a disutility
attaches to work is not the highest that could be attained.

We think this is usefully suggestive.   A common observation in
less developed countries is the apparent low utilization of labour,
even when the marginal product of labour is positive, indeed
rather high in some cases.    Certainly, the causes of this phenomenon are many and complex.    However, our model shows that
the true trade-offs between work and consumption are not likely
to be found by comparing the productivity of work with the
extra energy required in the short-term and the disutility of
work.    The implications of a given work pattern in the long-
term for long-term energy requirements are an essential feature
of the problem.


REFERENCE

Bliss, C.J. & Stern, N.H. 1978, Productivity, Wages and
Nutrition:   Part I, The Theory and Part II, Some Observations, Journal of Development Economics, 5, 331-62 and
363-98.

# TIME AND AFFECT: FACETS OF THE SOCIAL ANTHROPOLOGY OF WORK

Sandra Wallman
Resource Options Programme,
London School of Economics,
Houghton Street, London WC2

## 1. INTRODUCTION

Dialogue between academic disciplines around a recognisably common topic is likely to broaden the mind and challenge the assumptions of even the intellectual conservatives among us. But if it is stimulating, it is also hazardous. The hazard lurks in the communication procedure: insofar as the other discipline's formulations are unfamiliar or esoteric, they will sound pointless because meaningless, or brilliant because incomprehensible – the assessment probably depending on some previous and well established personal stance. Either way, the other person's work can be misunderstood. On top of this, two special difficulties arise when the interdisciplinary communication takes the form of the social scientist talking to natural or 'real' scientists. The first is obvious: the social scientist cannot hold other things equal, tends to use too many ifs and buts and too little (if any) statistical verification, and often does not come across as 'scientific' in the popular sense at all. The second special difficulty was signalled in a statement made by T.H. Marshall decades ago. He observed that good social science is not a matter of making people say "My goodness, how extraordinary!", rather is it a matter of making them say "Ah yes. Now I remember." Social science should, by this token, sound familiar to ordinary social creatures, even to professional scientists - although its very familiarity is likely to confirm them in the view that social science is not scientific. On this basis, even if this contribution fails to meet the formal standards of a biological symposium, it can be said to have succeeded to the extent that it reminds the reader of things he already knows about ordinary life.

This chapter has three parts. Part I reports themes raised in the 1979 Conference of the Association of Social Anthropologists, convened (by the author) around the topic 'Work'. Papers presented at that conference included accounts of pre-

industrial, industrializing and late industrial systems. Two
emphases emerged from the discussion: (a) Similar dimensions
of work could be identified in widely different systems, although
in comparative perspective they were shown to have very different significance: (b) the notion of work could not usefully be
limited to matters of physical effort and material production.

Part 2 explores themes from the social anthropology conference which have special bearing on this Symposium in human
biology. These are that the products of Energy and the incentives to Effort are governed generally by other things happening,
other options, other elements of the context in which they are
expended; and, more particularly, the effects of Energy and
Effort on the actor/worker and on his society vary according to
the criteria by which they are evaluated, and the extent to
which the worker is identified with the goals, the product or
the status of the work.

Part 3 describes two field-research strategies developed in
the study of resource options and resource management in inner
London. 'Maps' on which informants' investments of value and
identity in the time they spend and the people they know are
shown, and the possible relevance of those investments to the
computation of energy and effort is suggested.


2.    PART 1

Both the A.S.A. Conference and the volume arising out of
it (Wallman 1979b) were called Social Anthropology of Work,
but neither purports to cover the social anthropology of work.
Nor does this paper: there is not (yet) a designated area of
enquiry or theory which defines the topic in a way that justifies
using the definite article. The social anthropologists involved
in the 'work' Conference represented several different approaches
to the topic; and even the handful taking part in this Symposium can only be lumped together by contrast to other disciplines
represented. But social anthropologists can agree to certain
observations of the present world. One is that 'work' is an
increasingly topical issue: not only are local, general and
international elections fought over employment and the 'right to
work' on the one hand, and matters of resource allocation and
personal and regional autonomy on the other, but these items
are now somehow bracketed together in the popular mind. The
other is that the meaning and even the definition of 'work' is
changing throughout industrial society and also, perhaps by
contagion, in the nonindustrial world, but none of us knows
how to assess the changes.

The editorial framework imposed on the conference papers was articulated around two questions. The first: <u>What is work about?</u> is taken up in Part 2. The second: <u>What does social anthropology have to say about work that has not already been said?</u> can be answered by reference to three perspectives characteristic of the discipline.

(a)    <u>Social anthropology is defined by its comparative pers-pective.</u>  It rests on the assumption that not everyone class-ifies and evaluates the same world in the same way, nor do they do so consistently all the time.  Given this assumption we might expect energy and effort to mean different things in different cultures or different circumstances;  and we might wish to con-sider how far any differences of meaning will affect the physio-logical or biological processes of the human organism.

(b)    <u>Social anthropologists pay particular attention to con-text.</u>  When seeking to understand a social item or a set of events, they are trained to ask:  What else is happening? What else could be happening?  How much of other things happening 'counts' and in what way?  Thus:  if a man is labouring on one occasion alongside someone he likes... has on another just had a quarrel with his wife... is planning a sermon or a political speech... is expecting rain... is teaching his son the task... the social anthropologist recognizes that these five energy-spend-ing events are different from each other, even if the same amount of calories is consumed in every case.

(c)    <u>Social anthropologists normally speak of the 'whole' of social life as a system made up of interlocking parts or subsys-tems.</u>  Some find it useful to visualize these subsystems as domains or spheres – whether of activity, of exchange, or of meaning – to which particular resources pertain, and between which they can, in the right circumstances, be transposed. The model is of course only heuristic, but it can account for incon-sistencies of meaning, behaviour or value in a given social system (Barth 1963, 1966, 1967, Paine 1974).
These three perspectives were implicit in the 'work' confer-ence.  The fourteen papers given were presented under four headings:  <u>Systems of Organization and Authority;</u> <u>Systems of Value and Values;</u> <u>Systems of Identity;</u> and <u>Concepts of Work.</u> Perhaps inevitably, however, most of the papers turned out to be 'about' all these things and a fixed classification of their con-tent obscured more than it revealed.  The editorial framework introducing them for publication was built instead around a shopping list of the dimensions of work revealed in the separate papers.  The list included energy, incentive, resources, value,

time, place, person, technology, identity and the systematic
relations among them.    The point made was that each system
contains many constituent elements, and each element reacts or
interacts with every other in some way, but their relative importance varies:    the separate dimensions of work do not count in
the same way in every system, nor are they ranked in any fixed
order of priority by the participants in it.


3.    PART 2

    In this section the complex discussion of the A.S.A. Confer-
ence is summarized as six informal propositions answering the
question 'What is work about?'    Emphasis is put on the similarities and differences between 'work' and the human biology of
energy and effort whenever possible.


3.1.    Proposition 1

    Work is 'about' the application of human energy to resources:
this application converts, maintains, or adds value to the worker,
the thing worked on, and the system in which the work is performed.    This physical definition can deal with the work of
economic sectors and occupational groups as much as with individual workers.    It could even be made to take in the work of
managers, brokers, intermediaries, artists, ritual specialists
- all those whose work cannot be measured by its material product or 'use value'.    In physiological terms, each is expending
energy and is therefore altered by the work effort;    even psy-
chic energy burns calories.    And if 'the thing worked on' can
be read to include non-material resources (institutions, symbols,
information), then all systems of work are energy systems.    But
from the social-anthropological perspective, work can never be
understood as a mechanical function of energy expenditures:
human work is deliberate effort, directed energy;    it is energy-
directed to more or less explicit goals.
    These goals can be generalized as the performance of neces-
sary tasks and the production of necessary values - moral as
well as economic.    The task of meeting obligations, securing
identity, status and structure, are as fundamental to livelihood
as bread and shelter.    On this basis, work may be defined as
the production, management or conversion of the resources
necessary to livelihood - 'the sum total of capital, skills and
social claims' (Frankenburg re Barth 1966).    These will vary

according to environment and technology and, of course, according to what is regarded as 'necessary'.    But for the moment, let us assume the necessary resources to be six:    the classical trio, Land, Labour and Capital (or their equivalents) are joined by Time, Information and Identity.    Each can be used for or converted into the other.    Work is then not only about the production of material goods, money transactions and the need to grow food and to cook the family dinner.    It must equally be about the ownership and circulation of information, the playing of roles, the symbolic affirmation of personal significance and group identity – and the relationship of each of these to the other.

## 3.2.    Proposition 2

Because it is <u>deliberate</u> effort, <u>directed</u> energy, <u>Work is 'about'</u> <u>incentives</u>.    In Tikopia as in business economics, an element of purpose is contained in the simplest definition of work:    work is the expenditure of human energy to accomplish ends, normally with some sacrifice of comfort and leisure (Firth 1979).    But as soon as we ask what is the effort *for*, it becomes clear that the reasons for working vary.    Even in the formal job market and despite the heat generated by wage bargaining in democratic industrial systems, few would seriously argue that the maximization of money or other material gain is the sole form of rational economic behaviour or the only incentive to effort.    By the same token, work is not only 'economic' activity:    it "is at one time an economic, political and religious act, and is experienced as such" (Godelier 1972, 266).

The point is plain once it is illustrated.    In this society, strike action is as often about the dignity of working conditions as it is about wage rates, and some people go into factory work 'only' because it is more sociable than being in an office or at home, or because they want to be with their friends (see e.g., Klein 1976, Beynon & Blackburn 1972, Saifullah Khan 1979). The incentives of industrial workers are not unlike those of Melanesian gardeners in this respect (Schwimmer 1979).    Certainly, the most serious implications of redundancy and unemployment in industrial processes are not the narrowly economic.    When the state (or the extended family) pays the basic bills and the absence of a job has no monetary implications, the 'work' of maintaining social status and personal esteem is most often reported as paramount, whether the focus is on authority within the family or credibility in the community (R. Harrison 1976, Marsden & Duff 1975, Fagin 1979, Wadel 1973, 1979).    Sometimes the incentive to regain formal employment is hard to distinguish

from a pressing need to put structure and purpose back into
everyday life, i.e., to direct energy, to justify effort.
    The incentive to take up a particular kind of work or even
to work at all has a logic distinct from incentives to working
well.    Most obviously, people are inspired to make the extra
effort necessary for a better performance by the promise of extra reward.    It need not of course be material reward:    in industrial society, personal satisfaction is said to be better than
money in the bank.    In money economies as well as in societies
where money values are beside the point – either because there
is no currency or because money is used only in very limited
spheres of life – the incentive to a specially good performance
or product may be its extra value in use or in direct exchange,
but it is probably as often the kudos or 'differentials' gained
from special skill, or the satisfaction of having created something with which one is quite simply proud to be identified.
    Questions about incentive are usually of the kind: how can
people be persuaded to make the effort necessary to accomplish
particular ends?    what gives them the incentive to work especially well at particular tasks and to take pride in the work they
do?    how is work to be rewarded when subsistence is anyway
assured – whether because nature provides warmth and fish and
fruit, or because a reasonable minimum wage is guaranteed by
an affluent state?    But the really interesting question turns
the problem of incentives around:    how come people tend to
work as much or as little when they are not forced to, as when
they are?


3.3.    Proposition  3

    Work is 'about' control – physical and psychological, social
and symbolic.    The primaeval purpose of work is the human
need to control nature, to wrest a living from it and to impose
culture on it.    By definition there is no social group that does
not 'work' at this level.    Because the business of controlling
nature is a matter of technical ingenuity, it is both eased and
complicated by inventions of culture and patterns of organization.
The control of nature is therefore transposed into a more general
need to control the environment and the business of livelihood.
Central to the rubric of the A.S.A. conference was the fact that
the working relationship between man and nature is never un-
embroidered;    and that much of the social-cultural embroidery
on work tends to be concerned with the control of one person or
category of people over another – whether direct control by
means of command over the actions of others, or indirect control

achieved either by limiting their access to resources and benefits (Nadel 1957), or by devaluing the resources and benefits
which they have.

In the loosest possible sense, people can be said to 'make'
others work just as they 'work' machines, but the analogy is
limited by three particular considerations.   Firstly, work is
'about' social transaction as much as material production.  Indeed,
its significance to the worker lies more often in the quality of
the relationships involved than in the bald facts of material
survival (Firth 1967).

Secondly, work controls the identity as much as the economy
of the worker, whether as an individual or as the member of a
caste or an occupational group.    Being (also) a psychological
matter, it is both more and less than economic activity;    however instrumental or impersonal the attitude of others to his
work, it is, for the worker, a personal experience, his relation to the reality in which he lives.   At the same time it is
only a tiny part of the wider processes of the environment and
must be understood in an economic framework which constrains
the choices, decisions and rewards of the worker.

Thirdly, the control of work entails not only control over the
allocation and disposition of resources, it implies also control over
the values ascribed to each of them.    Evaluation of the energies
and efforts of others is more subtle than the physical control of
their actions, but it affects their livelihood no less importantly.
We need to therefore not only ask what forms of effort are called 'work' in each setting and why; we need also to know which
forms of work are in that setting considered socially worthy and
personally fulfilling.


## 3.4.    Proposition  4

Work is 'about' the production, management and conversion of
value as much as of material resources.   The value of energy
or of effort may be assessed in social, personal, or material terms
and measured by moral or economic criteria.    Nor are the weights
of these values fixed.    The relative value of any resource depends
on what it is being measured against, its converted value on the
technical processes applied to it, and its added or 'surplus value'
on the structure and state of the market.    None of these value
dimensions is autonomous:    in any one system, the value of particular forms of effort, even of work itself, depends on other
elements in that system.    This is as true for individuals as for
social forms:    the extent to which a person values one kind of
effort above another depends not only on the values of the society

in which he lives, but on other things happening to him or her at the time – other options, other constraints, other obligations – and the same expenditure of physical energy has different values in different domains of livelihood.    These differences are readily observed in ordinary life.    We may assume, for example, that the caloric value of digging so many square metres of garden is constant – at least for one individual.    But its economic value changes if I dig to grow potatoes instead of ornamental shrubs;  and its moral value is drastically different if, having no horticultural purpose at all, I am digging to bury stolen goods or a dead alley cat.    Similar shifts can occur on the level of relationships:    the moral, social, existential etc., value of the effort I put into relating to my own spouse changes if I apply the same effort to relating to someone else's partner.

Whatever the central measure of work, the assessment of efficiency, productivity and worthiness must depend on where, in any system, the actor sits.    This is most obviously crucial because structural or status position governs the resources at his/her disposal.    It is also crucial because the social assessment of energy values is an assessment of relative worth.    The value that I put upon my work is in some part a function of what else I could be doing;  what I see others doing;  what I expected to be doing at or by this time;  what I hope to do next year;  whether my position has changed and whether that change is for the better  (Wallman 1977, 1979a);  whether I am prepared to sweat for the sake of my own or my children's future benefit, or I must have my returns now (compare, e.g., Kosmin 1979 with Willis 1977).    Migrants in any system are said to work harder and to put up with conditions that no indigenous worker would tolerate exactly because they tend to be future-oriented and to see conditions 'away', however inhospitable, as somehow better than conditions at home (Murray 1979, Parkin 1979, Wallman 1979a).    All these values are governed by other things happening or, more directly, by what particular categories of people know about or think about what is happening.    The value of energy is therefore a function of its opportunity costs  (what am I giving up to achieve this end?)  and of its alternative costs (how else could I achieve it?  could anyone else do it for me?)  In each case it is necessary to be clear whose evaluation is at issue and in which context;  and to consider what kinds of value are being produced or converted by the energy and effort of the worker.

## 3.5.   Proposition 5

Work is 'about' personal identity and identity investment.
Contradictions in the value of effort and energy are tolerable
to the extent that contexts or domains of valuation can be held
distinct and the degree of personal investment in each controlled.
In this way, economically worthless work can be personally
highly valued, and socially despised tasks can be a source of
pride and identity (Cohen 1979, Searle-Chatterjee 1979);   the
fact that works of art and ritual performances are neither useful
nor negotiable in exchange does not preclude their being of
immense personal or social worth.     Contradictions of this kind
are so commonplace that the extent of discrepancy between economic and moral 'gain' can be used as a criterion for the comparison of systems of work (Schwimmer 1980).
     The worker may resolve anomalies of his/her own position
by identifying with only one dimension of work.     Work may,
for example, be defined in terms of the place in which it is
done.     Thus:   Melanesians prefer to work 'for nothing' in the
village to doing the same work for money outside (Schwimmer
1979);   cleaning one's own house is unpaid work but may en-
hance the reputation of the cleaner, while cleaning someone
else's house is rewarded in money but not in self-esteem.     In
other circumstances the worker ignores specific dimensions and
defines himself by the lifestyle associated with a particular job.
Sometimes he has no option:   the efforts of white-collar workers
to preserve differentials of salary and patterns of consumption
are made unusually desperate by the fact that their occupations
have no other locus of identity (Fred 1979).
     In every case the individual can take the option of identify-
ing with one domain of livelihood by minimizing or denying his
investment in others;   and he can shift his identity investment
from one domain to the other according to the opportunities and
constraints of circumstance.     The shift is probably easiest in
'complex' industrial society, where the domains of livelihood are
curiously discrete, but it is not always painless.     Each domain
needs attention (occupational status is not 'ascribed' it must be
'achieved';   marriage must be 'worked at';   political concern
should be 'demonstrated') and each contributes to the compo-
site identity structure of the individual.     Too narrow a focus
of effort or energy therefore creates other problems.     Where
'over-identification' occurs it is diagnosed as pathology:     anyone too closely identified with one domain of work is a 'work-
aholic', bound to be neglecting other obligations and probably
suffering from stress.     The healthy balance would seem to be
a spread of effort and energy across all of livelihood so that

each kind of work gets and gives its due.

The reverse of identification with work or aspects of the work process is alienation from it.    Just as complete identification is unlikely, so is total alienation.    Logically, it would mean the worker being – not just being treated like – a machine. Degrees of alienation are greatest where the worker has negligible control over the value and the disposition of his product, least where he initiates the work effort, organizes his time and energy and can identify with the product and the values of the product.

Some systems compensate the worker for bad conditions or low status by paying him disproportionately high wages or allocating him special areas of autonomy – i.e., a negative value in the personal sphere is recompensed by a positive economic good – although neither money nor power seem to reduce the physiological stress entailed by alienation as effectively as does satisfactory identity in another domain    (compare Harrison 1979 and Loudon 1979 with Searle-Chaterjee 1979).


### 3.6.    Proposition 6

Work is 'about' symbolic and productive investments in time. Obviously, the time values of human effort vary according to the technology available:    tools extend the efficiency or the energy of the worker.    A given piece of work done by hand costs so much time and energy (although these costs vary with strength and skill), but if a machine is bought to do the same task, (some) human energy will be replaced by machine energy and the input of time reduced.

Less obviously, the time value of work will vary according to when it is done.    'Overtime' can be paid at the value of 'time-and-a-half' only if it has been agreed that a specific quantity of time is worth more at particular moments in the daily, weekly, or annual cycles.    On a different time dimension, personal age or 'time of life' is pertinent too, in two respects.    One is that some tasks become easier with experience, others harder with age:    it was observed in the A.S.A. conference discussion that activities designated pleasure and play in youth may turn, in time, into laborious obligations.    The other is that the perception of amounts of time and so of time cost is not consistent throughout a lifetime:    a young child feels the year between Christmases like a century, but his grandmother insists that 'the days drag and the years fly'. The difference is due in some part to changes in biological processes, but it must also reflect patterns of livelihood.    In

some cultures, at least in the performance of some tasks, no
time cost is computed - i.e., time, as such, appears to have no
value (Wallman 1965, 1969, Schwimmer 1979). From this perspective the scope for using time as a measure of value in economic spheres without a money currency has been very limited
(see Belshaw 1954, Firth 1979, Minge-Klevana 1980). Even
where money values pertain, time and money often belong to
different work equations: work structures time; money rewards (some) work. Remember the unemployed man in Marienthal in the 1930s whose day had no time structure and so was
'empty'. Remember too that he reported talking to his wife
as an obligation, along with waking his children for school,
fetching wood and eating a midday meal (Jahoda et al. 1972).
It is as though, in the absence of formal employment, other
activities took up the function of giving shape and purpose to
his day - they 'became work' - although not in this case satisfying work. We could argue that they already were 'work'
on the grounds that they were necessary to livelihood, but while
the value of even the most essential tasks remains unacknowledged,
'hidden', those tasks will not be dignified by the status of work
and their performance will score no points, either social or existential, for the worker (Wadel 1979). We might also ask how and
how far the various activities of livelihood can be computed along
a single time dimension. Does less time spent at one mean more
time available for another? How elastic is the time curve?
Does the unemployed man at home in fact spend more time tending the garden or relating to his wife and children than he did
when he had a job which kept him out of the house all day?

Levels of energy govern both the perception and the productivity of time. These are not strictly a matter of kilojoules.
Consider: when my partner is away for days or weeks I have
many more household tasks to perform. Logically, I must be
'spending' his share of task time as well as my own. But if
he is absent I can spend no time at the work of relating to him.
The number of working hours in the day is not changed, but
the balance of the day's work may be altered - more household
tasks, less affective energy. Do I therefore spend more or
less energy than before? Do I have more or less to spend?
At what point is the lack of emotional recharge reflected in
overt energy levels?

Ideally, if we want to know the relation between time and
work we should know not only who does which necessary tasks
and how long each takes, but also who chats to whom, how
much time each parent spends (or thinks it spends) with its
partner, with each child. While X was doing the cooking
or washing the dishes, what else was happening? Was there
someone else at home who should have been doing the job?

What were they doing instead?    It is not hard to ask an informant who plants the yams and who tends the garden, and to
observe roughly how long each task takes.    But we need to
know also who else <u>could</u> have done it, and whether there was
anyone else in the garden who had nothing to do with the hoe-
ing but worked at relating to the gardener.

Work in a sociable atmosphere or work which, while accom-
plishing economically necessary tasks, also fulfils <u>social</u> obligations should be distinguished from work which has no explicit
social dimension.

## 4.    PART 3

These propositions are convincing enough in logic and familiar in common sense, but are they useful in research?    How
much can they be made to add to our understanding of the work
of livelihood?    How far (and this, indeed, is the proper measure of scientific <u>and</u> social scientific quality) can they be tested?
This section begins with a brief account of research into the
Resource Options of different population categories living in
the same inner-London neighbourhood.    It then describes fieldwork strategies exploring variations in the ways in which informants manage the <u>time</u> and the <u>people</u> resources available to
them.    In the terms of this Symposium, these strategies are
designed to allow us to assess and compare the <u>affective energy</u>
invested in particular tasks, relationships or domains of livelihood.

The Resource Options Programme has both a specific and a
general relevance to current social policy concerns.    On the
first count it will explore aspects of the integration of minority
groups into ordinary life in London.    Two questions are fundamental:    how far do people of different origins living in the
same area, and apparently with access to the same formal resources, behave differently in the management of livelihood;
and, in what circumstances do people of different background
identify themselves in terms of that difference?    On the
second count, it raises the more general question:    what makes
for viability in contemporary urban life?    On both counts the
real issue is:    how come some households manage and others
– superficially quite similar – do not?

This study pursues issues raised in the Social Anthropology
of Work Conference described in Part 1:    it is concerned with
work and the significance of work which falls outside the scope
of formal economic analysis, and with the elaboration of research
strategies by which it can be monitored.    The research is

interdisciplinary. It will build on the Ethnicity Programme
(SSRC-RUER 1975-1980) which concentrated on variations and
changes in the use of ethnicity as a principle of economic organization and group identity – the two strands of 'work' designated – in one specific inner London area, expanding the
analytic context to take account of other principles by which
localized populations may organize or identify for particular
purposes, and extending the scope for generalization by making
a controlled comparison of the local area of the first programme
with one to be selected for the second.

The study involves a comparison of two residential areas
– one in each of two inner-city boroughs – on the basis of
historical records and a full ethnographic survey (see Wallman
et al. 1980). This is followed by detailed case studies comparing the ways in which selected households perform and experience the work of livelihood. The strategies reported
below are aspects of this second stage of the enquiry: they
constitute part of the effort to analyse these households as
systems of resources, to map the domains or spheres of livelihood to which the various resources pertain, and to achieve a
better understanding of the prerequisites to 'viability'.

Because the household resource systems which constitute the
second stage of the enquiry are to be compared with each other,
they are selected out of the neighbourhood ethnographic survey
to have certain characteristics in common. Five criteria are
used in the selection. Each household (a) has lived in
the area for at least five years; (b) has kin living in the
area but outside the household; (c) has children under
sixteen living at home; (d) falls into the same socio-economic
status category; and (e) has experienced a normal crisis
(a birth, a death, a move, a redundancy etc.) in the preceding
twelve months.

Criteria (a) and (b) ensure that all the households
studied have the same possibility of involvement in the local
area and the option of calling on locally-based kin for assistance
or support; criterion (c) puts them at a similar stage in
the domestic life cycle; (d) confines them to a range of
skilled and semi-skilled, manual and non-manual occupations (in
OPCS categories 3 and 4) which are both non-executive and
weekly paid. Each of these criteria controls, to some extent,
the resources available to each household and the 'work' over
which those resources must be deployed; while criterion (e)
provides a focus for discussion, 'what-did-you-do-when...?',
of a situation requiring extra attention to resource options
and resource management.

The information contained in the time and network maps
shown here is integrated with other levels of data. Most signi-

ficant are the demographic, sociographic and geographic 'facts' about each household revealed in the first-stage neighbourhood survey and the detailed job histories also collected in the second stage.    Some levels of the analytic effort are less specific but no less significant to the comparison of households:    we have found it particularly useful to pay attention to those orientations and expectations covered by the general notion of 'philosophy of life'.

The designs of figure 2 (prepared by David Clark) and figure 4 (adapted for these purposes by Andra Goldman) allow a number of inferences concerning the affective energy invested by each household in its <u>time</u> and <u>people</u> resources to be drawn.

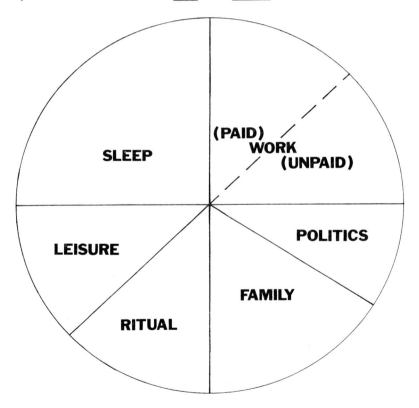

Figure I        Time/energy schedule 'A':   designed to show a 24 hour cycle of separate and distinct tasks.

These inferences are made plainer by contrast to figures 1 and 3 which record only the formal dimensions of the resource at issue, and ignore the context which decides both its meaning for the actors and its relation to other domains of livelihood. Figures 5 and 6 are versions of figure 4, filled in by two separate households. The contrast between them indicates quite different patterns of affective energy expenditure. Figure 1 represents a conventional time/work schedule. It shows energy expenditure in terms of crude proportions of time spent on categories of activity or in particular domains of livelihood. Changes in the designation of any category will not lift the one-dimensional constraint.

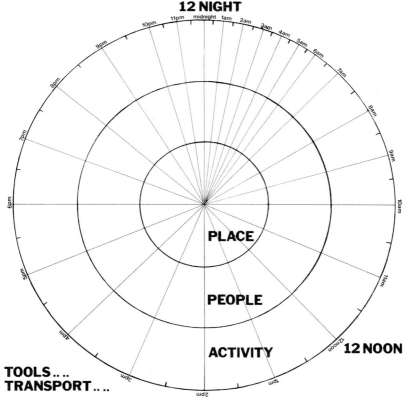

Figure 2     Time/energy schedule 'B': designed to show other things happening through each task period.

Figure 2,    by contrast, shows energy expenditure in a number
of dimensions.    It records time spent in designated activities
along with other things happening, i.e., in relation to the people,
the tools, the places and the technology involved.
    'Other things happening' can be significant to the energy
spent over a specific period of time in two ways.    If those
things are also 'work', they too must be consuming caloric energy;
and because the meaning of any task is altered by its context,
so the worker's affective energy investment in it will change
with the logic of the situation in which it is performed.    Questions as to whether these two kinds of energy are analogous,
whether they should be or can be computed together, whether
affective energy is amenable to physiological measures of any
kind – these are only questions of biological method and technique.    The difficulties of measurement which we may encounter
as scientists or social scientists do not alter the realities of
human livelihood.
    Figure 3 represents the conventional form of people/distance
schedule.    It allows the informant (as an individual or a household) to record significant others (i.e., actual or potential
people resources) in terms of geographic distance:    how close
are you to each of them in fact?    Figure 4, on the other hand,
places the same significant others in terms of their relative affective distance from the informant(s):    how close are you to
each of them in feeling?    Most informants find the second
classification no more difficult than the first;    some, on the
contrary, tackle it with more enthusiasm.    The triangular section marked off in both versions for antipathetic or ambiguous
relationships does sometimes give them pause, but none has denied that such relationships are 'extra' in some way (see
further Wallman et al. 1980).    In the framework of this chapter,
they can be said to represent an 'extra' drain on affective
energy.    Both maps are used in the present study since one
of its concerns is the function of localism as a principle of organization or identity in the two research areas:    the extent
to which people who are affectively important to the respondents
are also geographically close is an indicator of the latters' investment or involvement in the local area.
    Finally, figures 4 and 5 show the affective networks drawn
by two different households in the same area.    The contrast
between them is marked in terms of the absolute number of
people recognized as important in each case, and in the quality
or closeness of the people resources they designate.    The
real meaning of these differences does not emerge from the maps
alone, but a number of observations are suggestive.    Figure
5, for example, shows a total absence of entries in ring 2
which represent individuals who have been brought into the
household by one of its members and, while remaining

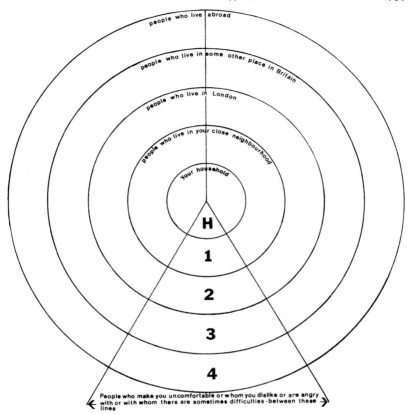

Figure 3          People/distance schedule 'A':    the geographic
network.

important to that member, are now also important to or close
to the household as a whole.    The same map indicates that,
of a reasonable number of kin recognized and named (ring 4),
rather few are important for particular purposes (ring 3) or
held in specially close esteem (ring 1).    Similarly, on the
non-kin side only three people are consistently or specially
close (ring 1), but quite large numbers are important for some
purposes (ring 3) or recognized and named but not often 'used'
as resources (ring 4).    Figure 6 shows quite a different balance, with relatively more people in the inner rings on both
sides, but the outermost ring unusually loaded in the 'tension

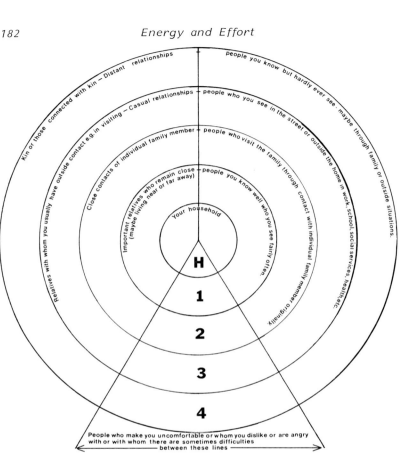

'Figure 4          People/distance schedule 'B':     the affective network.

zone'.
     Clearly, these two households are organizing their
affective social lives differently.    When these differences are
placed in the context of their separate cultures, expectations
and economic circumstances, the material adds a crucial perspective on the two households as resource systems.     It also
allows us to consider the systematic comparison of the affective
energy put into particular domains of livelihood by households
which share the same formal environment but manage the resources available to them in different ways and sometimes to
crucially different effect.

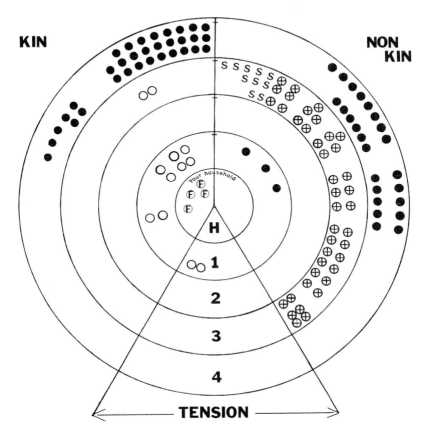

Figure 5      Affective network:      household 'X'.

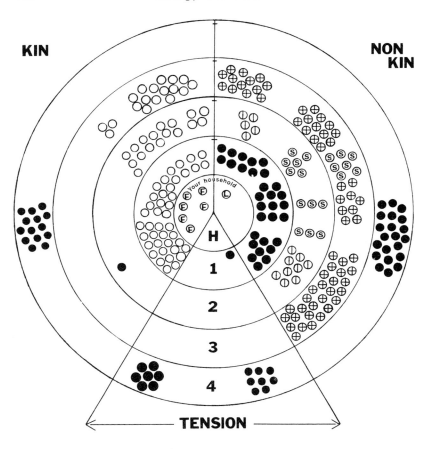

Figure 6      Affective network:      household 'Y'.

# REFERENCES

Barth, F., 1963, Introduction to The Role of the Entrepreneur in Northern Norway (Oslo: Scandinavian University Books).

Barth, F., 1966, Models of Social Organization, Occasional Papers No 23 (London: Royal Anthropological Institute).

Barth, F., 1967, Economic Spheres in Darfur. In Themes in Economic Anthropology, edited by R. Firth, Association of Social Anthropologists Monograph No. 6 (London: Tavistock).

Belshaw, C., 1954, Changing Melanesia (Melbourne: OUP).

Beynon, H. & Blackburn, R.M., 1972, Perceptions of Work: Variations within a Factory. Cambridge Papers in Sociology No. 3 (CUP).

Cohen, A.P., 1979, The Whalsay Croft: traditional work and customary identity in modern times. In Social Anthropology of Work, edited by S. Wallman, Association of Social Anthropologists Monograph No. 19 (London: Academic Press).

Fagin, L., 1979, Views from three other disciplines: psychiatry. In Social Anthropology of Work, edited by S. Wallman, Association of Social Anthropologists Monograph No. 19 (London: Academic Press).

Firth, R. (editor) 1967, Themes in Economic Anthropology, Association of Social Anthropologists No. 6 (London: Tavistock).

Firth, R., 1979, work and value: reflections on ideas of Karl Marx. In Social Anthropology of Work, edited by S. Wallman, Association of Social Anthropologists Monograph No. 19 (London: Academic Press).

Frankenburg, R., 1966, British community studies: problems of synthesis. In The Social Anthropology of Complex Societies edited by M. Banton, Association of Social Anthropologists Monograph No. 4 (London: Tavistock).

Fred, M., 1979, How Sweden Works: a case from the bureaucracy. In Social Anthropology of Work, edited by S. Wallman, Association of Social Anthropologists Monograph No. 19 (London: Academic Press).

Godelier, M., 1972, Rationality and Irrationality in Economics (London: NLB).

Harrison, G.A., 1979, Views from three other disciplines: biological anthropology. In Social Anthropology of Work, edited by S. Wallman, Association of Social Anthropologists, Monograph No. 19 (London: Academic Press).

Harrison, R., 1976, The demoralising experience of prolonged unemployment. Department of Employment Gazette, London: April, 339—48.

Jahoda, M., Lazarsfeld, P.F. & Zeisel, H., 1972, Marienthal: the Sociography of an Unemployed Community, (Cambridge: CUP).

Klein, L., 1976, New Forms of Work Organisation (Cambridge: CUP).

Kosmin, B.A., 1979, Exclusion and opportunity: traditions of work among British Jews. In Ethnicity at Work, edited by S. Wallman (London: Macmillan).

Loudon, J.B., 1979, Workers, Lords and Masters: the organization of labour on South African farms. In Social Anthropology of Work, edited by S. Wallman, Association of Social Anthropologists Monograph No. 19 (London: Academic Press).

Marsden, D. & Duff, E., 1975, Workless: Some Unemployed Men and their Families (Harmondsworth: Penguin).

Minge-Klevana, W., 1980, Does labour time decrease with industrialization? Current Anthropology, 21, 1.

Murray, C., 1979, The Work of Men, Women and the Ancestors: social reproduction in the periphery of Southern Africa. In Social Anthropology of Work, edited by S. Wallman, Association of Social Anthropologists Monograph No. 19 (London: Academic Press).

Nadel, S.F., 1957, The Theory of Social Structure (London: Cohen and West).

Paine, R., 1974, Second Thoughts about Barth's Models, Occasional Paper No. 321 (London: Royal Anthropological Institute).

Parkin, D., 1979, The Categorization of Work: cases from coastal Kenya. In Social Anthropology of Work, edited by S. Wallman, Association of Social Anthropologists Monograph No. 19 (London: Academic Press).

Saifullah Khan, V., 1979, Work and network: South Asian women in South London. In Ethnicity at Work, edited by S. Wallman (London: Macmillan).

Schwimmer, E., 1979, The Self and the Product: concepts of work in comparative perspective. In Social Anthropology of Work, edited by S. Wallman, Association of Social Anthropologists Monograph No. 19 (London: Academic Press).

Schwimmer, E., 1980, The Limits of Economic Ideology: a comparative study of work concepts.    International social Science Journal, 33.

Searl-Chatterjee, M., 1979,   The polluted identity of work: a study of Benares Sweepers.    In Social Anthropology of Work, edited by S. Wallman, Association of Social Anthropologists Monograph No. 19 (London:   Academic Press).

Wadel, C., 1973, "Now, whose fault is that?" the struggle for self-esteem in the face of chronic unemployment    (St. Johns, Newfoundland:   Memorial University ISER).

Wadel, C., 1979, The hidden work of everyday life.    In Social Anthropology of Work,   edited by S. Wallman, Association of Social Anthropologists Monograph No. 19 (London:   Academic Press).

Wallman, S., 1965, The communication of measurement in Basutoland.    Human Organization, 24,   236—43.

Wallman, S., 1969, Take Out Hunger:   two case studies of rural development in Basutoland,    LSE Monographs No. 39 (London: Athlone).

Wallman, S. (editor), 1977, Perceptions of Development, (Cambridge:   CUP).

Wallman, S. (editor), 1979a, Ethnicity at Work (London: Macmillan).

Wallman, S. (editor), 1979b, Social Anthropology of Work.   Association of Social Anthropologists Monograph No. 19   (London: Academic Press).

Wallman, S., Dhooge, Y., Goldman, A. & Kosmin, B.A., 1980, Ethnography by Proxy:   strategies for research in the Inner City.    Ethnos   45, 1—2.

Willis, P., 1977, Learning to Labour (Farnborough:   Saxon House).

# RESOURCE DISTRIBUTION AND FORAGING EFFORT IN HUNTER-GATHERER SUBSISTENCE

D.R. Harris
Department of Human Environment,
Institute of Archaeology,
University of London,
31 Gordon Square, London WC1

## 1.    INTRODUCTION

It is now over a decade since the once-dominant view of hunter-gatherer life as "nasty, brutish and short" (Lee 1968) was replaced by a new orthodoxy, summed up by Sahlins (1968) when he described hunter-gatherers as "the original affluent society".    This paradigm shift stemmed mainly from field studies of contemporary hunter-gatherers in Africa (Lee 1969, Turnbull 1968, Woodburn 1968) and northern Australia· (Gould 1969, Mc-Arthur 1960, McCarthy & McArthur 1960) and it was forcefully articulated by Sahlins in his book Stone Age Economics (1972). The essence of this new view was that hunting and gathering provided a secure livelihood and afforded ample time for leisure. Far from eking out a precarious living searching unremittingly for food, the hunter-gatherer groups studied were shown to spend on average only 2—5 hours per adult per day obtaining and processing foods.    (However, Hawkes & O'Connell 1981 have recently challenged this generalization and pointed out, with particular reference to Lee's data on the !Kung San, that if full allowance is made for the time devoted to food processing, this average appears to be a substantial underestimate.)    They enjoyed varied diets and appeared to live well within the carrying capacities of the environments that they occupied.
The first, and most obvious, objection to this highly generalized reassessment of hunter-gatherer life is that it is based primarily on studies of a small sample of groups living today in tropical desert, rain-forest, and savanna environments.    By combing the ethnographic literature, however, Sahlins was able to show that historically known hunter-gatherers in a wide variety of physical environments, from the temperate forests of eastern North America to the subantarctic coasts of Tierra del Fuego, secured plentiful supplies of food with relatively little effort.    A second and more powerful criticism of the new orthodoxy is that the modern and historical descriptions of hunter-

gatherer groups, on which Sahlins' analysis depends, represent
snap-shots in time.      They tell us nothing about the long-term
relationships of these groups to their environments.      How representative such groups really are of hunter-gatherers in general, both worldwide and through time, is therefore still an open
question.      I do not deny that Lee's and Sahlins' reassessment
has provided a much-needed corrective to earlier ill-founded
assumptions about the precariousness of hunter-gatherer subsistence, but here I wish to advance the debate by asking two
basic questions.      First, on what attributes of physical environments should we focus attention in analysing subsistence opportunities and constraints, and second, how do the hunter-gatherers themselves respond to those opportunities and constraints by
means of different subsistence strategies?      In this chapter,
therefore, I examine both the environmental arena in which the
action takes place, and the subsistence-related activities of the
actors.


## 2.    THE ENVIRONMENTAL ARENA

An analytical framework that is often adopted in generalized
discussions of hunter-gatherer groups is that of major terrestrial ecosystems or biomes.      This yields highly simplified categories, such as hunters of the Arctic tundra and temperate
grasslands, temperate coastal fisher-folk, desert foragers, and
tropical rain forest collectors.      It is the approach that Forde
adopted in the early 1930s when he was writing Habitat, Economy
and Society (1934).      It still has some pedagogic value at an
introductory level, but because each biome is characterized so
broadly - usually in terms of simple environmental parameters
such as average annual or monthly temperature and rainfall,
predominance of forest, grassland, or desert vegetation etc. -
it fails to specify with any precision which attributes of the
physical environment most directly affect the subsistence behaviour of hunter-gatherers.
A somewhat more refined approach to the overall ecological
context of hunter-gatherer subsistence is to analyse biomes
in terms of gradients of variation of their biotic and abiotic
characteristics.      Thus, if attention is focused on, for example,
floral and faunal diversity, as one significant variable, it is
possible to classify biomes according to their degree of biotic
specialization.      By this means, highly specialized biomes, in
which species diversity is low but each species is represented
by many individuals, as in the Arctic tundra and the boreal
coniferous forest, can be contrasted with much less specialized

(i.e. 'generalized') biomes, such as the equatorial evergreen rain forest, in which species diversity is very high and each species is represented by relatively few individuals.   This mode of analysis allows general statements to be made about differences in the availability of food resources to hunter-gatherers, particularly in terms of contrasts in spatial distribution:   a point that is further discussed below.

Another variable by which biomes can usefully be classified and compared along a gradient of variation is net primary productivity.   However, because this is simply a measure of the average amount of plant material available in an ecosystem for consumption by herbivores, omnivores and decomposer organisms, it tends to be too all-inclusive a parameter to be of direct relevance to the analysis of hunter-gatherer subsistence.   It can, however, be applied to those plants – and, by assessing secondary productivity, to those animals – in an ecosystem that are known to be exploited for food by man, and thus yield a significant measure of the potential capacity of that ecosystem to support a human population, assuming a given technology of food procurement (Foley 1977).   While this appears to offer a promising method for assessing the subsistence opportunities for hunter-gatherers afforded by different biomes, in practice its application is severely limited by a lack of field data on primary, and especially secondary, productivity in many ecosystems.   For the forseeable future, it is likely that attempts to characterize and compare the physical environments of hunter-gatherers on a world scale will continue to be based largely on abiotic variables, particularly temperature and precipitation, for which adequate data exist.   That such variables can, however, be used in a sophisticated way as surrogates for direct measures of biotic productivity has recently been demonstrated by Binford (1980) in an ingenious analysis that relates effective temperature, which describes both the annual amount and the seasonal variation of solar radiation, to a sample of 168 hunter-gatherer groups classified according to their degree of residential mobility.

We can now return to the theme of contrasts in the spatial distribution of food resources between different types of ecosystem, with reference to the concept of 'fine-grained' as opposed to 'coarse-grained', 'patchy', or 'clumped' patterns of distribution.   This distribution has, in recent years, been profitably applied by biologists to studies of the feeding behaviour of various species of birds, insects and mammals. Like other dichotomizing concepts, it represents as two categories of resource-distribution pattern, or, more generally, of environment, what is in reality a gradient or continuum from the most fine-grained or uniform to the most coarse-grained or diverse

pattern.    It does, nevertheless, allow us to make useful generalizations about resource availability.    Thus, in fine-grained
environments, such as the equatorial rain forests, where species
diversity is high and individuals of the same species tend to be
spatially separated, wild food resources available for human exploitation are distributed more randomly than they are in
coarse-grained environments, such as many desert and polar
ecosystems, where species diversity is lower and individuals of
the same species tend to be spatially concentrated.

Such contrasts in spatial distribution also have a temporal
dimension in terms of when, in their reproductive cycles, species
are most readily or profitably exploited by the human population.
Seasonal cycles in the availability of plant and animal foods are
the most obvious expression of temporal—spatial variation in
resource availability, but shorter-term (e.g., diurnal) and
longer-term (e.g., inter-annual) periodicities may also be significant.    Non-periodic or episodic events, such as volcanic
eruptions, earthquakes, tsunamis, and hurricanes of exceptional
strength, also, of course, affect the availability of resources
locally, but their very unpredictability makes it very difficult
to incorporate them into generalized models of resource availability and hunter-gatherer behaviour.

Although the concept of fine-grained and coarse-grained environments can be applied at the highly generalized level of
major world biomes, it is more realistic to consider its implications for resource availability at a regional scale more consistent
with the areas occupied by individual hunter-gatherer groups.
Empirical observations of hunter-gatherers have repeatedly demonstrated that they prefer to locate their camps at or in the
vicinity of ecotones between contrasted ecosystems:    at forest
and woodland edges, along upland-lowland margins, and
at such conjunctions of land and water as river banks, swamp
fringes, and lake and sea shores.    Around such locations the
overall distribution of resources is necessarily coarse-grained
or patchy, although within any one local ecosystem a fine-
grained pattern may prevail.    Thus, at the regional scale,
defined in terms of the area consistently occupied and exploited
by a hunter-gatherer population, patchy patterns of resource
availability appear to be the norm and fine-grained patterns the
exception.    Even hunter-gatherers living wholly within areas of
tropical rain forest tend to locate their camps with respect to
ecotones within the forest, particularly along river and stream
courses (Harris 1978).

The advantages of location in coarse-grained environments
include not only access to a wide range of types of resource,
but also a greater probability of predicting the spatial-temporal
occurrence of resources.    Clearly, the less random the occur-

rence of resources in a given environment the higher their predictability.    Whereas in a tropical forest it is likely to be necessary to walk considerable distances to find more than one or two individuals of a particular food-yielding species, in coarser-grained ecosystems the chances of encountering clusters of such species at known locations and seasons are much higher.

Application of the concept of relative predictability is not restricted to different types of ecosystem.    It can also be applied to differences in the types of resource exploited.    The most elementary distinction to be made is that between plants and animals.    Because of their capacity for locomotion, animals are inherently less spatially predictable than plants, although for many animals aspects of feeding, mating, nesting, and migratory behaviour are sufficiently localized in space and time to be readily predicted.    Obvious examples of such behaviour are the regular drinking at the same dry-season water-hole that is characteristic of many game animals in tropical savanna environments, or the long-distance seasonal movements of such strongly migratory species as salmon, caribou and reindeer. Detailed analysis of the predictability of patterns of movement of prey animals in a given ecosystem or mosaic of ecosystems can thus provide a useful measure of one set of resource opportunities open to hunter-gatherers.

In analysing the available plant resources in terms of their predictability, attention focuses particularly on their lifespan. The location of long-lived perennial plants, such as nut- and fruit-bearing trees, can - given adequate knowledge of the local area - be predicted precisely, and the timing of their fruiting seasons tends also to be highly predictable, although this may vary considerably from species to species and even from tree to tree:    for example, in the summer-green deciduous forest biome of eastern North America, in Illinois, hickory trees (Carya spp.) produce nuts in a relatively regular mast cycle of 2—3 years, whereas the mast cycles of oaks (Quercus spp.) are generally longer and more irregular (Asch et al. 1972). The occurrence of shorter-lived perennial plants, especially those that are herbaceous, and which in markedly seasonal environments experience die-back of their above-ground parts during dry and/or cold periods, is less easily predicted.    One category of herbaceous perennials, which makes a major contribution to many hunter-gatherer diets, consists of those plants that develop large underground food-storage organs in the form of roots, tubers, rhizomes, corms and bulbs.    Insofar as their dry and withered leaves and stems may be difficult to find in forest, woodland, or thicket at the time when they are due to be harvested, their precise location can be said to be unpredictable, but they tend to persist in the same local habitat from

year to year and can usually be found by reference to some
more permanent and conspicuous feature, such as a large tree,
in the vicinity.   Yams constitute a particularly important tub-
erous food resource for tropical hunter-gatherers, and the location of wild yam grounds is often recorded by leaving a
marker of some kind.   In northeastern Queensland, for example
Aboriginal people may break off a branch and place it in the
thicket where the yam vines grew (Harris 1977).   It must
also be remembered that some tuber-yielding plants, notably
manioc (cassava), are shrubs rather than herbs and have
woody stems that persist above ground from year to year. The
least predictable categories of plants are annuals and ephemerals,
which only survive from one generation to another as dormant
seeds.   They include annual grasses and forbs, many of which
produce edible seeds of relatively high nutritional value that
are  exploited by hunter-gatherers, and their general areas of
occurrence are likely to be approximately known from year to
year in all environments except those, such as very dry deserts,
that are subject to extreme inter-annual fluctuations of rainfall
and runoff.

   In sum, therefore, it may be argued that the physical-environ-
ment arena of hunter-gatherer subsistence is best analysed not
in terms of biome type, biotic diversity and productivity, or
abiotic variables, such as rainfall and temperature, but in terms
of patchiness and predictability in the occurrence of individual
resources within a given ecosystem or mosaic of ecosystems.
However, techniques for analysing these two critical attributes
in the field, and for presenting them in ways that allow precise
comparisons to be made from one area or subsistence system
to another, remain to be worked out.


3.   SUBSISTENCE BEHAVIOUR

   Having examined - at least superficially - the environmental
component, we can now reverse the coin and look at subsistence
more explicitly from the point of view of the hunter-gatherers
themselves.   In doing so, I shall make use of some of the concepts developed by animal biologists working on models of optimal foraging strategies, or what MacArthur calls "the economics
of consumer choice" (1972).   As MacArthur points out, the
gathering of food by a foraging animal involves four phases:
the decision where to search, the search itself, the decision
whether to pursue a potential item of food once it has been found,
and the pursuit itself;   actions that are then followed by the
possible capture and eating of the food item.   Translated into

a vocabulary more suited to hunter-gatherer behaviour, we can reword and elaborate these phases as follows:

(a)    the decision as to which patch or patches within the exploitation area will be visited on a given occasion;
(b)    movement to and within the patch(es) to be exploited in order to locate a food resource;
(c)    the decision whether to exploit a particular resource once it has been located;
(d)    exploitation of the resource.

This last phase is divisible into (a) the procurement (collection or capture) of food, (b) its processing, and (c) its consumption, and a particular aspect of food processing that constitutes a significant variable in hunter-gatherer subsistence is whether it incorporates techniques of storage, which have the capacity to extend greatly the time interval between procurement and consumption (cf. the distinctions drawn by Binford (1980) between "collectors" who practise food storage and "foragers" who do not, by Woodburn (1980) between hunter-gatherer economic systems of "delayed return" and of "immediate return", and by Testart (1981) between mobile hunter-gatherers who do not depend on food storage and sedentary hunter-gatherers and agriculturalists who do so on a large scale; see also Ingold's (1980) dissection of the concept of food storage into its "ecological", "practical" and "social" manifestations).
    Before looking more closely at the question of choice in, and energy expended on, the food quest, it is necessary to state the basic assumption on which models of optimal foraging rest, i.e., that the feeders maximize energy return relative to the time invested in the food quest.    It follows from this assumption that hunter-gatherers will tend to forage where the expectation of yield, relative to the time invested, is greatest, an expectation which will vary from place to place and from time to time in relation to the spatial-temporal patchiness and predictability of resource distribution.    The second element of choice integral to the food quest is the decision whether to exploit a resource once it has been located.    At that juncture the forager can either collect (or pursue) the plant or animal, or search for a preferred, higher-ranked food instead.    If the latter decision is taken, more time will be expended in the expectation of a higher (more satisfying) return relative to the time invested.    In other words, we can say that when foragers locate a food resource they will only exploit that resource if the net return from it exceeds the return likely to be gained from searching for and exploiting another, higher-ranked resource.

If we further assume that the overall ranking of food resources relates more directly to energy gained per unit of time
expended on their search and exploitation than it does to other
criteria, such as their nutritional value or rarity, then we
should expect that the importance of low-ranked resources in
the diet will vary inversely with the abundance – or more
accurately the accessibility – of high-ranked resources.   Thus,
if access to high-ranked resources increases, low-ranked resources will tend to diminish in importance or even be eliminated
from the diet, and vice-versa, regardless of the overall abundance
of the low-ranked resources (Schoener 1971).   This inference
has interesting implications for understanding both spatial variations and temporal changes in hunter-gatherer susbsistence. In
particular, it obliges us to recognize that the status accorded to
a resource by hunter-gatherers can vary widely from one context to another, dependent on what other resources are available to them.


4.    HUNTER-GATHERERS AS OPTIMAL FORAGERS?

    In order to test whether the subsistence behaviour of hunter-
gatherers accords with the predictions of optimal use of food
resources outlined in the previous section, we should now
examine a range of case studies drawn from differing environmental contexts.   Unfortunately, this cannot be done, because
we almost completely lack detailed and comprehensive investigations of time expended and energy gained by hunter-gatherers
in the food quest.    A notable exception to this generalization
is, however, O'Connell & Hawkes' (1981) analysis of plant
use by the Alyawara Aboriginal people of the central Australian
desert, which attempts explicitly to apply a model of optimal
diet to hunter-gatherer resource use.    In the remainder of
the chapter, I summarize the main findings of that innovative
study, comment briefly on some less well documented examples
from other parts of Australia, and conclude with some specu-
lations about further implications of optimal foraging models for
the understanding of hunter-gatherer subsistence.
    The starting point of O'Connell and Hawkes' analysis is
the observation that, although the Alyawara traditionally collected edible seeds from nearly half of the 85 plant species in
the subsistence inventory, in recent decades when they gained
access to European foods, particularly flour, they ceased to
exploit wild seeds systematically despite the abundance and
accessibility of this resource, and despite the fact that other
wild-plant foods continue to be gathered on a small scale.    The

question to ask is, therefore, why have the Alyawara stopped eating the seeds of those wild grasses and trees which had been staple foods in the traditional diet and which continue to be locally abundant today? O'Connell and Hawkes argue that this question is best answered in terms of the energetic costs and benefits of exploiting various resources; more specifically, because wild seeds are energetically expensive resources to gather and process they will be exploited only when the returns from other resources are low, regardless of the absolute abundance of the seed-yielding plants.

The traditional homeland of the Alyawara is located about 250 kilometres north-east of Alice Springs, Northern Australia, in desert country consisting of grass-covered sandy plain, dune fields, and low sandstone ridges intersected by normally dry stream channels, with scattered patches of acacia-dominated scrub woodland (mulga). The establishment of European cattle stations in the area in recent decades has changed the life-style of the Alyawara from one of high mobility, attuned to the seasonal availability of food and water, to one in which the people live in semi-permanent settlements near the cattle stations, where they have become heavily dependent on government welfare distributed in the form of weekly rations of flour, sugar, powdered milk and tea. Households with a cash income also buy additional foods (sweet biscuits, canned fruit and meat, fresh beef and soft drinks etc.) at station stores. Men still hunt kangaroos, euros and bustards, and thus provide up to 70% of the total per capita meat intake, and women still collect wild foods, although this activity now accounts for less than 5% of total food intake. O'Connell and Hawkes present data on wild-plant collecting carried out by the Alyawara from the settlement of Bendaijerum (MacDonald Downs) in 1974-75. This activity took place in three major plant communities or habitats: mulga woodland, riverine floodplain and sandhills. Quantitative data were obtained for a total of 19 collecting trips, or some 12—15% of the estimated 100 to 125 trips that took place from Bendaijerum between May 1974 and March 1975.

Although the sample is small, there is no reason to think that it is unrepresentative of contemporary plant-collecting activities overall. The most striking observations to emerge from the analysis are (a) that the Alyawara collected fewer than 12 plant species out of the 85 that they identify as edible, (b) that only three of these (Ipomoea costata, Vigna lanceolata and Solanum centrale) were regularly collected in any appreciable quantity, and (c) that only one species (Acacia coriacea) was collected by choice for its (almost ripe) seeds, despite the fact that 46% of the species in the traditional inventory produce edible seeds. The failure to exploit seeds cannot be explained

TABLE 1.   Net energy returns and resource ranking for some wild plants and animal foods procured in 1974–75 by the Alyawara people of Northern Australia (modified from O'Connell & Hawkes 1981).

| Habitat and resource | Handling time 1 (h/kg) | Mean caloric value (kcal/kg) | Net caloric return 2 (kcal/h) | Rank order |
|---|---|---|---|---|
| Sandhill | | | | |
| Ipomoea costata | 0.25 | 1563 | 6252 | 1 |
| Solanum centrale | 0.50 | 2992 | 5984 | 2 |
| Acacia coriacea (unripe) | 0.60 | 2600 | 4333 | 3 |
| Varanus sp. | 0.25 | 1050 | 4200 | 4 |
| Vigna lanceolata | 0.50 | 862 | 1724 | 5 |
| Acacia coriacea (ripe) | > 5.25 | 3551 | > 676 | |
| Acacia aneura | 6.50 | 3778 | 581 | |
| Grass seeds | 6.00 | 3450 | 575 | 6 |
| Acacia cowleana | 6.50 | 3589 | 552 | |
| Other acacias | 6.50 | 3500 | 538 | |

| Habitat and resource | Handling time 1 (h/kg) | Mean caloric value (kcal/kg) | Net caloric return 2 (kcal/h) | Rank order |
|---|---|---|---|---|
| **Mulga woodland** | | | | |
| Amphibolarus sp. | 0.25 | 1050 | 4200 | 1 |
| Varanus sp. | 0.50 | 862 | 1724 | 2 |
| Vigna lanceolata | 1.75 | 2600 | 1486 | 3 |
| Cossid larvae | 6.50 | 3778 | 581 | |
| Acacia aneura | 6.00 | 3450 | 575 | 4 |
| Grass seeds | | | | |
| **Riverine floodplain** | | | | |
| Cyperus sp. | 0.75 | 3326 | 4435 | 1 |
| Grass seeds | 6.00 | 3450 | 575 | 2 |

1. Handling time = collecting + processing time per person per kilogram of the resource.
2. Return per person per hour of handling time once the resource has been encountered.

in terms of lack of availability or access, and it is unconvincing
to argue that the paste prepared from wild seeds is less pala-
table than the roots of Vigna or Ipomoea.    Why then were the
seeds ignored (with the exception of those of Acacia coriacea)
in favour of less abundant and generally more distant roots,
fruits and witchitty grubs (cossid larvae)?

In trying to answer this question in terms of a model of
optimal diet, O'Connell and Hawkes calculate the relative energy
return per unit of handling (collecting and processing) time for
each resource obtained in the three habitats, and rank the
resources accordingly (table 1).    On 11 visits to the sandhills,
energy returns ranged from about 750 to 5200 kilocalories (kcal)
per forager hour searching, collecting and processing, and
returns per unit of handling time for each resource that was
taken in quantity – Ipomoea roots, Solanum fruits and the
nearly ripe seeds of Acacia coriacea – were well above the average
return from the habitat.    Although ripe seeds of several species
were available in the habitat and were often encountered, they
were not collected because the returns from other, higher-ranked
resources never fell low enough to make seed collection worthwhile.
Returns from ten visits to the mulga woodland were lower than
those from the sandhills, ranging from about 200—800 kcal/for-
aging hour.    There, the main resources exploited were lizards,
witchitty grubs and Vigna roots, all of which yield returns in
excess of 1400 kcal/hour (table 1), and again no ripe seeds
were collected.    Only one trip to the riverine floodplain was
recorded and so the data obtained are not strictly comparable
with those from the other two habitats.    It is probably significant, however, that only Cyperus corms were gathered and
that, although grass seed was available on the floodplain, none
was collected, which in view of the high estimated handling time
for grass seeds, accords with expectation (table 1).

The results from all three habitats are thus seen to be generally consistent with the predictions of foraging behaviour derived
from a model of optimal diet.    The model also helps to explain
why seeds were so important in the traditional, pre-European
diet of the Alyawara.    In the absence of European foods, the
exploitative pressure on native plant-food resources would have
been greater and the probability of seeds forming part of the
optimal diet would have been higher, even though their exploitation would have necessitated the energetically costly manufacture
and maintenance of seed-grinding tools.    O'Connell and Hawkes
point out that tools for processing seeds are common in household
debris at an abandoned campsite 30 km northwest of Bendaijerum,
which was occupied in 1971—73 at the end of a long period of
drought.    The effects of prolonged drought on the productivity
of local sandhill and mulga habitats, coupled with reduced access

to European foods in this more remote location, was evidently sufficient in that instance to make seed exploitation worthwhile there at that time, in contrast to the wetter and more productive seasons of 1974—75 in the foraging areas nearer Bendaijerum.

O'Connell's and Hawkes' study of wild-plant exploitation by the Alyawara is the most explicit and detailed application of the optimal-diet model to Australian hunter-gatherers, but several other studies of hunter-gatherer subsistence in northern tropical Australia tend to support the predictions derived from the model. The most comprehensive and highly quantified of these is the investigation of subsistence among the Anbarra group of the Gidjingali people of the Arnhemland coast that has been carried out in recent years by Jones and Meehan (Jones 1980, Meehan 1975, 1977a,b). Like the Alyawara, the Anbarra have access to flour, sugar, tea and other European foods, and today most of their carbohydrate intake is derived from these foods; but wild-plant foods are still collected, and animal foods are obtained by hunting, fishing and gathering. In 1972—73 the caloric contribution of European flour and sugar to the diet varied from a maximum of 57.8% in the late wet/early dry season to a minimum of 35.2% in the late dry season, while 40—50% of the average intake of calories came from animal flesh, fish contributing about 30% of this intake, terrestrial mammals about 11% and reptiles about 7% (Jones 1980). The Anbarra themselves maintain that this balance of vegetable and animal foods corresponds closely to that which prevailed in their traditional diet before they had access to European foods, when the major sources of carbohydrate were the tubers of wild yams (Dioscorea transversa and D. bulbifera), the leached seeds of cycads (Cycas media) and the corms of the spike rush (Eleocharis dulcis). The occurrence of these wild foods is highly predictable spatially and temporally, but they are time-consuming to collect and process, particularly the cycads which require detoxification, and it is not surprising that they have been largely superseded by European sources of carbohydrate. If the yields of various wild foods hunted, fished and gathered by the Anbarra (expressed in kcal per person per hour on a successful trip) are compared it is found that all animal foods except bivalve shellfish outrank the two plant foods - yams and cycads - for which data are available (table 2). The continued dependence of the Anbarra on high-ranking animal foods, and the decline in their exploitation of the low-ranking plants that were former staples, is thus seen broadly to fit the optimal-diet model, although the data presented by Jones for animals and plants are not strictly comparable to those recorded for plants only by O'Connell and Hawkes, so that a quantitatively equivalent rank order cannot be calculated for the Anbarra. It is nevertheless a striking fact that the buffalo

TABLE 2. Net energy returns for some wild plant and animal foods procured in 1972–73 by members of the Anbarra community of the Gidjingali people of Arnhemland, Northern Australia (modified from Jones 1980).

| Resource | Mode of procurement | Mean caloric value (kcal/kg) | Net caloric return (kcal/person/h) |
|---|---|---|---|
| Water buffalo (Bubalus bubalis) | Hunted | 3000 | 200 000 |
| Agile allaby (Macropus agilis) | Hunted | 3000 | 50 000 |
| Fish | Speared | 1400 | 35 000 |
| Fish | Trapped | 1400 | 35 000 |
| Fish | Hook & line | 1400 | 6000 |
| Yam tubers (Dioscorea spp.) | Dug | 1300 | 1300 |
| Cycad seeds (Cycas media) | Collected & leached | 2000 | 1300 |
| Bivalve shellfish | Gathered | 800 | 1000 |

(Bubalus bubalis), which was introduced from Indonesia in the first half of the nineteenth century and now exists in large feral herds in Arnhemland (Calaby 1980), and which has, since the early 1970s, become the most highly prized quarry of An-barra hunters, is also by far the highest-yielding resource available to them.

Another example of contemporary adjustment to changes in resource availability that appears to fit the optimum-diet model derives from work by Chase and myself among the Aboriginal people of the Lockhart River community on the eastern coast of the Cape York Peninsula, Queensland (Chase in the press, Chase & Sutton 1981, Harris 1977). Although detailed quan-titative data on foraging effort and yields are not available, the response of the Lockhart River people to the availability of European foods is essentially similar to that of the Gidjingali in Arnhemland. At Lockhart, access to flour, sugar, tea and other European foods resulted in the dropping from the foraging repertoire of low-ranked wild foods, such as the 'bitter' varieties of wild yam (Dioscorea spp.), the germinating embryos of certain species of mangrove (e.g., Bruguiera gymnorhiza), and the seeds and stems of several rain-forest vines and trees (e.g., Entada scandens, Archontophoenix alexandrae, Gulubia costata). The common denominator of these resources is that because of their initial unpalatability (in some case their actual toxicity) they have to be processed in laborious ways before they can be consumed. They are abundant in the area and their occurrence is spatially and seasonally predictable, but they have been discarded from the diet while other, less labour-intensive and more highly valued wild foods such as 'sweet' yams, honey, green turtles (Chelonia mydas) and dugongs (Dugong dugon), continue to be procured. In particular, the dugong, a large, high-yielding marine mammal, occupies a posi-tion in contemporary resource valuation comparable to that of the buffalo among the Anbarra, and like the buffalo, its cap-ture confers on the successful hunter high social status as well as an abundance of calories.


5.    CONCLUSION

Lack of space and lack of data preclude further examination of the relevance of the optimum-diet model to contemporary Australian Aboriginal subsistence. In conclusion, it is impor-tant to stress that I do not imply that the principles of optimal foraging explain all aspects of subsistence choice and effort among hunter-gatherers. Other, non-energetic, social criteria undoubtedly influence subsistence behaviour.   Indeed, a major merit of the application of the optimal-diet model to hunter-gatherer subsistence is that it may be capable of revealing those choices that appear to be inconsistent with it and which therefore demand other types of explanation.   However, the examples of dietary changes that we have examined do seem to be consistent with the main prediction of the model, that in any given habitat, resources will be exploited, from the highest-ranked to the lowest-ranked, in order of the net return they yield per unit of time spent locating, procuring and processing them.

This conclusion suggests not only that the model helps to explain recent dietary changes, but also that it can contribute to our understanding of past patterns of hunter-gatherer subsistence.   In the conclusion to their paper on Alyawara plant use, O'Connell and Hawkes tentatively apply an optimal-foraging model to the prehistoric occupation of Australia by human populations.   They speculate that the order in which different types of environment were occupied prehistorically would have varied directly with the energy gained from their exploitation.   Thus, environments in which energetic returns were comparatively low, such as the central desert, would have remained unoccupied until returns in more rewarding environments, such as coastal and riverine areas, fell to the same level.   The archaeological record, scanty though it is, does not negate this interpretation, because on present evidence the central desert appears not to have been occupied until sometime after 15 000 B.P., whereas many coastal and riverine areas were occupied before 20 000 B.P. (Bowdler 1977, Gould 1977, Jones 1979).

Within any particular environment, too, we should, if hunter-gatherers are behaving as optimal foragers, and if the population is increasing and/or the resource base being depleted, expect to find evidence in the archaeological record for the early exploitation of high-ranked (low-cost/high-return) resources and the later exploitation of low-ranked (high-cost/low return) resources.   Perhaps this trend will prove discernible in the rain-forest region of northern Queensland if and when archaeological investigation provides evidence of the evolution there of the historically known subsistence system that focused on the exploitation of high-cost/low-return tree nuts in the context of relatively high human-population densities (Harris 1978).

Finally, and more generally, it can be suggested that long-term dietary trends from high-ranked to low-ranked resources may underlie the evolution of hunter-gatherer subsistence economies in many parts of the world, and that the insights afforded by optimal-foraging models may in particular help to explain the processes of transition from wild-food procurement by hunter-gatherers to food production by agriculturalists.

## 6.  ACKNOWLEDGEMENT

I wish to thank James O'Connell and Kristen Hawkes for generously allowing me to read (and to summarize in part here) their paper prior to its publication; it should also be pointed out that I have not had the opportunity to read any of the other contributions to the volume on hunter-gatherer foraging strategies in which their paper is to appear.  I wish to thank them also for reading and commenting on an earlier version of my paper.

## REFERENCES

Asch, N.B., Ford, R.I. & Asch, D.L., 1972, Paleoethnobotany of the Koster site: the Archaic horizons.  Illinois State Museum Reports of Investigations (Springfield, Illinois:  Illinois State Museum), 24, 1–34.

Binford, L.R., 1980, Willow smoke and dogs' tails:  hunter-gatherer settlement systems and archaeological site formation. American Antiquity, 45, 4–20.

Bowdler, S., 1977, The coastal colonization of Australia. In Sunda and Sahul: Prehistoric Studies in Southeast Asia, Melanesia and Australia, edited by J. Allen, J. Golson and R. Jones (London:  Academic Press), pp. 205–246.

Calaby, J., 1980, Ecology and human use of the Australian savanna environment.  In Human Ecology in Savanna Environments, edited by D.R. Harris (London:  Academic Press), pp. 321–337.

Chase, A.K., 1980, Which Way Now?  Tradition, Continuity and Change in a North Queensland Aboriginal Community (University of Queensland, Brisbane:  Unpublished Ph. D. thesis).

Chase, A.K., in press, Aborigines in Queensland: retaining identity in the face of decreasing options.  Paper presented at the Second International Conference on Hunter-Gatherers, Quebec City, September, 1980.

Chase, A.K. & Sutton, P., 1981, Hunter-gatherers in a rich environment:  Aboriginal coastal exploitation in Cape York Peninsula.  In Biogeography and Ecology of Australia, edited by A. Keast (The Hague:  W. Junk), pp. 1818-1852.

Foley, R., 1977, Space and energy:  a method for analysing habitat value and utilization in relation to archaeological sites. In Spatial Archaeology, edited by D.L. Clarke (London: Academic Press), pp. 163-187.

Forde, C.D., 1934, Habitat, Economy and Society (London: Methuen).

Gould, R.A., 1969, Subsistence behaviour among the Western Desert Aborigines of Australia.  Oceania, 39, 253-274.

Gould, R.A., 1977, Puntutjarpa rockshelter and the Australian Desert Culture.  Memoirs of the American Museum of Natural History, 54, 1-187.

Harris, D.R., 1977, Subsistence strategies across Torres Strait. In Sunda and Sahul: Prehistoric Studies in Southeast Asia, Melanesia and Australia, edited by J. Allen, J. Golson and R. Jones (London:  Academic Press), pp. 421-463.

Harris, D.R., 1978, Adaptation to a tropical rain-forest environment:  Aboriginal subsistence in northeastern Queensland.  In Human Behaviour and Adaptation, edited by N. Blurton-Jones and V. Reynolds (London:  Taylor and Francis), pp. 113-134.

Hawkes, K. & O'Connell, J.F., 1981, Affluent hunters? Some comments in the lights of the Alyawara case.  American Anthropologist, 83, 622-626.

Ingold, T., 1980, The significance of storage in hunting societies. Paper presented to the Troisième réunion "conservation des grains" Levrous, 24-29 November 1980.

Jones, R., 1979, The fifth continent:  problems concerning the human colonization of Australia.  Annual Review of Anthropology, 8, 445-466.

Jones, R., 1980, Hunters in the Australian coastal savanna. In Human Ecology in Savanna Environments, edited by D.R. Harris (London: Academic Press), pp. 107-146.

Lee, R.B., 1968, What hunters do for a living, or how to make out on scarce resources.  In Man the Hunter, edited by R.B. Lee and I. DeVore (Chigago:  Aldine), pp. 30-48.

Lee, R.B., 1969, Kung Bushman subsistence:  an input-output analysis.  In Environment and Cultural Behaviour, edited by A.P. Vayda (Garden City, N.Y.:  Natural History Press), pp. 47-79.

MacArthur, R.H., 1972, Geographical Ecology:  Patterns in the

Distribution of Species (New York: Harper & Row).

MacArthur, M., 1960, Food consumption and dietary levels of groups of Aborigines living on naturally occurring foods. In Records of the American-Australian Scientific Expedition to Arnhem Land, edited by C.P. Mountford (Melbourne: Melbourne University Press), Vol. II, pp. 90-135.

MacCarthy, F.D & McArthur, M., 1960, The food quest and the time factor in aboriginal economic life. In Records of the American-Australian Scientific Expedition to Arnhem Land, edited by C.P. Mountford (Melbourne: Melbourne University Press), Vol. II, pp. 145-194.

Meehan, B., 1975, Shell Bed to Shell Midden (Australian National University, Canberra: Unpublished Ph.D. thesis).

Meehan, B., 1977a, Hunters by the seashore. Journal of Human Evolution, 6, 363-370.

Meehan, B., 1977b, Man does not live by calories alone: the role of shellfish in a coastal cuisine. In Sunda and Sahul: Prehistoric Studies in Southeast Asia, Melanesia and Australia, edited by J. Allen, J. Golson and R. Jones (London: Academic Press), pp. 493-531.

O'Connell, J.F. & Hawkes, K., 1981, Alyawara plant use and optimal foraging theory. In Hunter-Gatherer Foraging Strategies: Ethnographic and Archeological Analyses, edited by B. Winterhalder and E. Smith (Chicago: University of Chicago Press).

Sahlins, M.D., 1968, Notes on the original affluent society. In Man the Hunter, edited by R.B. Lee and I. DeVore (Chicago: Aldine), pp. 85-89.

Sahlins, M.D., 1972, Stone Age Economics (Chicago: Aldine-Atherton).

Schoener, T.W., 1971, Theory of feeding strategies. Annual Review of Ecology and Systematics, 2, 369-404.

Testart, A., 1981, La conservation des produits végétaux chez les chasseurs-cueilleurs. In Les Techniques de Conservation des Grains à Long Terme, edited by M. Gast and F. Sigaut (Paris: CNRS), tome 2, pp. 181-193.

Turnbull, C.M., 1968, The importance of flux in two hunting societies. In Man the Hunter, edited by R.B. Lee and I. DeVore (Chicago: Aldine), pp. 132-137.

Woodburn, J., 1968, An introduction to Hadza ecology. In Man the Hunter, edited by R.B. Lee and I. DeVore (Chicago: Aldine), pp. 49-55.

Woodburn, J., 1980, Hunters and gatherers today and reconstruction of the past. In <u>Soviet and Western Anthropology</u>, edited by E. Gellner (London: Duckworth), pp. 95–117.

# QUALITY AND QUANTITY IN AGRICULTURAL WORK - SIERRA LEONE RICE-FARMING SYSTEMS

Paul Richards

Anthropology Department, University College,
Gower Street, London WC1E 6BT

## 1.  INTRODUCTION

Human biologists, interested in 'work', approach their
subject through attempts to quantify energy inputs and efforts
expended.  Social scientists, perhaps sceptical of the exaggerated claims made for energy-based input-output modelling techniques (cf. Burnham, this volume), argue that 'labour' is
more than physical effort.  A proper understanding of work
must embrace the study of labour as a social process.  Thus,
whereas the human biologist concentrates on physiological
measurements, the social scientist will tend to favour a 'linguistic' approach, decoding what informants say work means.
It is conventional to deplore the separateness of these two
perspectives, and to predict rich new insights should they be
combined.  This paper counsels caution.  Before such insights
can be obtained, a deep-seated problem affecting both approaches has to be overcome.  This is how to identify the vested
interests built into the data.  Descriptions of work processes,
it is argued, whether physiological or linguistic, quantitative
or qualitative, human-biological or social-scientific, tend to reflect the perspectives of specific groups.  Disinterested research
requires special safeguards.  The present paper illustrates this
theme, and concludes by suggesting the need to reject 'positivism' and to adopt an approach to energy and effort based on
dialogue, concerning research objectives, with groups to be
studied.

## 2.  AGRICULTURAL WORK IN SIERRA LEONE

Rice is the major staple of Sierra Leonian agriculture.
Traditionally, most rice has been produced on upland shifting
cultivated farms.  Most parts of the country receive over
2000 mm rainfall per year, and valley-bottom swampland is
widely available.  Agricultural development initiatives in the

last 30 years have concentrated on encouraging farmers to make more use of inland-valley swamps for rice cultivation. As urban populations have increased, rice imports have grown. Reducing the rice import bill is a major preoccupation of government. Swamp-rice yields per hectare are about 50% higher, on average, than yields on typical upland farms. It is thought that a shift away from use of rain-fed uplands towards swamp cultivation will help solve the urban food-supply problem at the same time as reducing ecological pressures caused by outmoded agricultural practices characteristic of shifting cultivators.

Results of schemes to encourage greater use of swampland for rice farming have been mixed. In the area around Makeni, in north-central Sierra Leone, the majority of farmers make some use of inland-valley swamps. An agricultural development project, the Northern Area Integrated Agricultural Development Project (IADP-North), jointly funded by the World Bank and the Government of Sierra Leone, has attempted to capitalize on this established interest by introducing farmers to 'improved' swamp-rice cultivation techniques, largely drawing upon Southeast Asian experience. Local swamp-farming practice is based around the notion of using the natural rise and fall of water in a valley-bottom swamp as the rainy season progresses. The improved practices of the IADP-North emphasize water control. A headbund and drainage channels are constructed, plots are levelled, and rice is transplanted. Local practice favours broadcasting seed over inundated land. The improved-practices package is labour-intensive, especially in terms of the initial constructional work required and the labour used in nursing and then transplanting rice seedlings, but in the official view rapid population growth has generated a surplus of rural labour. The problem of employing this 'surplus' labour is solved by adopting labour-intensive agricultural innovations. Within the sphere of operations of IADP-North, about 40% of male heads of farming households have adopted the project swamp-farming package (uptake by women farmers has been much less). Reactions vary. Some farmers are happy to continue with the new farming system, others are reverting to local swamp-management practices, or would prefer to shift back into upland rice farming (Karimu & Richards 1981).

Fieldwork in northern Sierra Leone has been complemented by detailed analysis of upland farming techniques in two villages south of Moyamba in southern Sierra Leone. Here, there has been no official scheme to encourage farmers to abandon upland shifting cultivation, nor have farmers shown more than a passing interest in swamp utilization according to local practice. Although inland-valley swampland is available, and some farmers have experimented with its use, informants vigorously

TABLE 1. Test for reliability of farmer recall data relating to farm size (source: fieldwork by Michael Johnny & Paul Richards, Moyamba District, 1978 & 1980).

| Farmer | Acres (measured) | 'Bushels' planted 1978 | 1980 Recall |
|---|---|---|---|
| 1. | 5.81 | 4.5 | 4.5 |
| 2. | 0.97 | 0.9 | 0.0 |
| 3. | 6.27 | 4.2 | 2.0 |
| 4. | 5.8 | 4.0 | 2.0 |
| 5. | No data | 12.0 | 12.0 |
| 6. | 10.21 | 9.0 | 3.0 |
| 7. | 2.40 | 3.5 | 2.0 |
| 8. | 2.87 | 2.8 | 3.0 |
| 9. | 3.98 | 2.5 | 2.0 |
| | $\bar{x}_1 = 4.79$ | $\bar{x}_2 = 4.8$ | $\bar{x}_3 = 3.4$ |

$\bar{x}_2/\bar{x}_1$ – 1 acre averages 1.22 bushels (allowing for missing value for farmer 5)

$\bar{x}_3/\bar{x}_2$ – recalls average 72% of actual 'bushels' planted

press the point of view that upland farming is 'better' (Johnny 1979, Johnny & Richards 1981).

The issue of swamp versus upland rice-farming systems in Sierra Leone provokes three categories of response, therefore: adoption of an 'improved' package of swamp-farming practices, maintenance of existing 'local' swamp-management techniques, and rejection of both types of swamp-farming system in favour of 'traditional' upland shifting cultivation. The purpose of this paper is to show that there is no single answer to the question of which is the best farming system. The interests of four groups are considered: government, wealthy trader-farmers, 'middle peasants' and women farmers. Each group carries out its own quantitative and qualitative analysis of appropriate 'returns to labour'. It is demonstrated that quan-tification is not disinterested. Farmers, as well as government agricultural economists, measure input-output relations, but in addition quantify aspects of returns to labour of little impor-tance to governments preoccupied with urban food-supply re-quirements. Furthermore, it is shown that informants don't just 'inform'; they construct. In a comparison of farm-size measurements and a small sample of farmers recalling these farm-size measurements 18 months later, the recall results were ± 25% of the measured figures, apart from the farmer with the second-largest farm in the sample, a chief, who under-reported its size by two-thirds (table 1). By the same token, moneylenders frequently represent themselves as poor. The paper thus argues the need to adopt a 'critical' approach, uncovering the links between 'knowledge' and 'human in-terests' (Habermas 1972), in studies which adopt either the quantitative 'input-output' approach of human biologists or the qualitative, 'discourse'-based approach of social scientists (Johnny et al., 1981, Richards & Sharpe (1981).

## 2.1. Quantification of returns to labour on upland rice farms

Figures in official sources such as the sample census of agriculture (Sierra Leone Central Statistics Office, 1970/1971) suggest upland rice-farm yields lie typically within the range 800-1200 kg/ha. Locally managed swamp-farm yields are per-haps 10-20% better, and, in principle, yields from swamps worked on 'improved' lines, with adequate water control, are 50-100% better. Even the latter figure may be exceeded if the swamp concerned has water flow throughout the dry season, thus permitting two crops a year. The expectation is, there-fore, that a vigorously pursued swamp-development policy will increase national marketable surpluses of rice by anything from

10-100%.  Labour inputs on upland farms are typically of the order of 180-220 days/ha (Spencer 1975, Johnny 1979).  Swamp farming is known to be more labour-intensive (see below). Various estimates for the annual labour requirements of improved swamp-farming systems made in the late 1960's and early 1970's suggest inputs/ha 25% greater than those typical in upland farming, but more recent figures suggest an increase in the order of 100% (Karimu & Richards 1981).  It now appears likely that swamp rice farming in many parts of Sierra Leone involves quite severe diminishing returns to labour investment, when compared with 'traditional' upland cultivation.  Nevertheless, this is not construed as a problem in official circles, because of the belief that rapid population increase in rural areas has created a labour surplus.

For the two case-study areas discussed in this paper, Johnny (1979) and Karimu & Richards (1981) suggest that this belief is ill-founded.  Because of selective outmigration of younger people, first of young males to Freetown and the diamond mining areas, and more recently of young females as well, net rates of population increase over the period 1963-1974 are small, and some chiefdoms have experienced absolute decline in numbers in the 15-60 age group.  Consequently, it is the case that labour is in short supply and expensive, especially in the area covered by IADP-North.  Much potential hired labour is available only through self-help labour co-operatives known as 'companies'.  A 'company's' first priority is to meet the labour needs of members.  Any unallocated labour time may be 'sold'. Members can acquire this 'surplus' labour at nominal rates. Rates for non-members are high - given prevailing prices for rice, about twice, per man-day, the amount the farmer can himself expect to earn.  The labour-intensive character of swamp farming is likely to exacerbate present labour shortages. Clearly, then, the official case for inland-valley swamp development, formulated in terms of data focusing on yields per unit of land, on the assumption that land is a scarcer factor of production than labour, is dictated more by political aspiration than economic realism.

Farmers also quantify their situation, even if their 'alternative' statistics are much less accessible to researchers than those in government publications.  Where farmers reject swamp farming for upland shifting cultivation, a critical factor in their decision is the greater scope offered by the upland farm for intercropping.  Gatherers of official statistics see upland farms as rice farms, and the presence of other crops as an inconvenience complicating the proper estimation of rice yields.  To the farmer, these additional crops provide a varied diet, and are vital to a family's chances of surviving the pre-harvest 'hungry

TABLE 2.   Average yields per hectare, intercropped upland rice farms Moyamba area, southern Sierra Leone (after Johnny 1979).

| Crop | | Yield kg/ha | Energy kJ x $10^6$ |
|---|---|---|---|
| Rice | Oryza sativa | 1063 | 8.94 |
| Sorghum | Sorghum margaritiferum | 200 | 1.73 |
| Fonio | Digitaria exilis | ? | ? |
| Belrush millet | Pennisetum leonis | ? | ? |
| Maize | Zea mays | 104 | 1.49 |
| Cassava | Manihot esculenta | 747 | 2.76 |
| Cassava leaves | | 95 | 0.16 |
| Sweet potatoes | Ipomoea batatas | 178 | 0.62 |
| Sweet potato leaves | | 64 | 0.11 |
| Yams | Dioscorea spp. | 87 | 0.22 |
| Cocoyams | Xanthosoma sagittifolium | 65 | 0.23 |
| Cucumber | Cucumis sativus | 261 | 0.18 |
| Pumpkin | Cucurbita spp. | 327 | 0.22 |
| Tomato | Lycopersicum esculentum | 149 | 0.13 |
| Garden eggs | Solanum melongena | 20 | 0.03 |
| Beans | Phaseolus lunatus | 297 | 3.40 |
| Egusi | Colocynthis citrullus | 68 | 0.11 |

| Crop | | Yield kg/ha | Energy kJ × 10⁶ |
|------|------|------|------|
| Pepper | Capsicum spp. | 9 | 0.01 |
| Beniseed | Sesamum spp. | 30 | 0.50 |
| Okra | Hibiscus esculentus | 57 | 0.10 |
| Krain–krain | Corchorus olitorius | 17 | 0.03 |
| Sawa–sawa | Hibiscus sabdariffa | 70 | 0.12 |
| Hondii | Amaranthus hybridus | 22 | 0.04 |
| | | | |
| TOTALS | | | |
| Rice | | 1063 (27%) | 8.94 (42%) |
| Other crops | | 2867 (73%) | 12.19 (58%) |
| Total | | 3930 | 21.13 |

season'. Although the 'hungry season' may be exacerbated by the kinds of environmental misfortune that spoil the rice crop from time to time, fieldwork suggests very strongly the conclusion that food shortages are closely linked to levels of indebtedness. Whether a poor harvest first caused indebtedness is of little relevance, since credit-debt relationships subsequently prove to be self-sustaining. The farmer pledges his crop to meet a food shortage and finds himself short of food the following year, irrespective of whether the environmental circumstances have improved or not. In this kind of situation, crops such as rice, cassava and maize, which can be substituted for the major staple in the weeks immediately prior to the rice harvest, are like money in the bank. Intercrops assume a strategic importance out of proportion to the area they occupy within the rice farm – not that this is small in any case.

The technical problems of land-use survey in intercropped farms are considerable. The approach adopted in the 1970 sample census of agriculture is to assume that intercrop yields are proportional to the amount of non-rice planting material mixed in with the seed rice at planting time, i.e., about 10–15% by weight. Yield-plot data for farms in the two villages of our southern Sierra Leone case study suggest that this assumption leads to a major underestimate of the importance of intercrops in the total output of an upland farm (Johnny 1979). An average intercropped farm has 22 food crops in addition to rice. This amounts to 75% by weight of the total output. A rough calorie count indicates that less than half of the total food output of the average upland 'rice' farm comes from rice (table 2). Rural-development strategies in Sierra Leone are not so much concerned with increasing food output as with converting an abundance of locally consumed intercrops into a marketable rice surplus for the benefit of urban consumers, and the data which underpin the official version of the rural-development problem, carefully compiled by 'scientific' research, reflect this vested interest. The apparent objectivity of these land-use survey procedures serves to disguise the political nature of current development initiatives.

Farmers are very conscious of the survival potential of an upland intercropped farm. Their own quantitative assessments are focused on this issue. (Yields per unit of area or time are of much less interest, since farm operations are organized on a task basis. If a work party is convened for, say, burning the farm, it is the group and the possibilities for sustaining it – by food, music, force of personality and political clout – which determines the ground covered and time spent. The Kantian notion that units of time and area have prior status as independent entities, from which a potential maximum productivity can

TABLE 3.  Reliability versus profit:  farmer preferences for low and high-yielding rice varieties.

| Preferred variety | Project farmers | | Non-project farmers | |
|---|---|---|---|---|
| | High yield/high risk variety [1] | Low yield/low risk variety [2] | High yield/high risk variety | Low yield/low risk variety |
| Matotoka heads of household | 4 | 15 | 1 | 18 |
| Bumban " | 15 | 3 | 12 | 4 |
| Gbendembu " | 3 | 17 | 0 | 19 |
| Matotoka women farmers | 0 | 3 | 0 | 19 |
| Bumban " | – | – | 10[3] | 7[3] |

1.  Hypothetical rice variety yielding 60 bn/ha and 20 bn/ha in good and bad years.  Good and bad years occur with equal probability.

2.  Hypothetical rice variety yielding 40 bn/ha and 30 bn/ha in good and bad years.  Good and bad years occur with equal probability.

3.  Groundnut cultivators.

be squeezed, is an alien concept.) Farmers regularly test the germination potential of batches of seed, and maintain 'trial plots' close to the farm hut for this purpose. The notion of experimentation – hungueh in Mende – extends to cover quantitative investigation of the yield potential of unfamiliar planting materials. A group of farmers was given seed new to them (improved upland varieties ROK 2 and ROK 3). No planting instructions – other than the information that the seeds were upland rice varieties – were given, but use of the material was carefully monitored. Several farmers organized their own input-output trials, marking out suitable plots and planting equal amounts of new seed and a familiar variety side by side. The utensil used to measure seed for planting would then be used to assess yields. Yield superiority in quantitative terms is an important factor in deciding whether to adopt or reject an unfamiliar rice variety. Relevant decisions were recorded at the time, and yield margins remembered in some detail 18 months later (Johnny 1979, Johnny & Richards 1981).

Farmers are not, however, solely concerned with quantitative assessments of yield advantages. Their experimental method has as its central concern the elucidation of interesting characteristics of planting materials in response to as wide a range of circumstances as possible. The upland farm is a jigsaw puzzle in which the complex interlocking of the pieces is the best possible guarantee of a stable output. The more farmers develop their knowledge of, and skill at handling, this complexity, the more satisfied they feel. Teasing out relevant information, whether quantitative or qualitative, is a constant preoccupation. Yet the existence of this huge fund of systematic knowledge is only rarely given its due by scientific bureaucracies. Ignoring, or denying, the experimental competence of the farming community is in effect to reject local problematics in favour of problem formulations dictated by outside interests.

Assessment of rice-seed characteristics is an especially good example. The 'outside' perspective is to concentrate on improving yield averages. On-farm evaluation of IADP-North improved rice varieties was conducted solely in terms of average values, ignoring variability of yield from plot to plot, or from year to year. And yet a sample of farmers from the project area expressed a marked preference for rice varieties which minimized inter-seasonal variability as opposed to maximizing long-term average yields. The results (table 3) show little difference between male and female farmers, farmers within and farmers outside the IADP North Project. The choice was rarely made on the basis of instinct or guesswork. Most farmers openly computed the trade-off between variability and profitability and explicitly chose risk minimization in contradistinction to profit

TABLE 4. Farmer preferences for swamp or upland rice farming systems, three sample settlements, northern Sierra Leone. (88 farmers*)

| | Farmer preferences for swamp or upland farming if restricted to only one type | |
| --- | --- | --- |
| | Swamp | Upland |
| Project farmers | | |
| Matotoka (n = 19) | 18 | 1 |
| Bumban (n = 18) | 18 | 0 |
| Gbendembu (n = 19) | 17 | 2 |
| | 53 (95%) | 3 (5%) |
| Non-project farmers | | |
| Matotoka (n = 19) | 12 | 7 |
| Bumban (n = 17) | 11 | 6 |
| Gbendembu (n = 19) | 12 | 7 |
| | 35 (64%) | 20 (36%) |

TABLE 5. Reasons cited for preferring swamp or upland farming systems.

Reasons for preferring swamp rice farming (88 farmers)

**The work is easier   60**

| | |
|---|---|
| requires less assistance/able to work alone | 24 |
| less time-consuming/daily attendance not essential | 10 |
| less pest control/bird scaring | 7 |
| less weeding | 5 |
| insufficient help from young people to farm upland/children at school | 8 |
| work is easier for women/old people | 5 |
| can be combined with off-farm work/less time-constrained/still obtain yield if start the work late | 5 |
| fewer separate operations in each year/work more routine/familiar | 5 |
| farm boundaries permanent (IADP farms) | 3 |

**Better rice yields   54**

| | |
|---|---|
| more profitable | 1 |
| more rice for same size of farm | 1 |
| more than one crop a year possible | 1 |
| unqualified | 51 |

Reasons for preferring upland rice farming   (23 farmers*)

| Intercropping | 23 | | |
|---|---|---|---|
| | | food even when rice is finished | 9 |
| | | greater variety of food, even if work is harder | 3 |
| | | more food/greater variety of food | 3 |
| | | more profitable (except in a bad year) | 5 |
| | | earlier rice harvest, and intercropping reduces chances of indebtedness | 2 |
| | | unqualified | 1 |
| Firewood sales | 8 | | |
| Familiarity | 3 | | |
| Upland rice tastes better | 1 | | |
| Swamp work is unhealthy for the old | 1 | | |

*Some farmers cited more than one basic reason, and provided several additional glosses.
Total of reasons cited is greater than the number of farmers interviewed.

maximization (Karimu & Richards 1981). Those preferring the profit-maximizing solution were for the most part wealthy trader-farmers. Despite the obvious conclusion that risk-avoidance is the ordinary farmer's first priority, variability analysis is still a poor second in the methodology of project appraisal (but for a worthy exception, see Dunsmore et al. 1976).

## 2.2.  Swamp versus upland rice farming - qualitative perspectives

Social scientists, working through local languages and able to decode subtle nuances of social meaning, will tend to approach the study of agricultural work not through a series of measurements but through their informants' descriptions. If the quantitatively oriented researcher's difficulty is to avoid over-reliance on measurements dictated exclusively by 'outside' interests, the social scientist's problem is to resist the temptation to treat informant's statements as representative of a general consensus. Sentences beginning 'Tiv say...' or 'The Tallensi believe...' are problematic because they disguise the fact that not all groups within society have the same power and opportunity to make appropriate programmatic statements to social researchers. If research into the social meaning and significance of agricultural work is to do more than represent the beliefs and attitudes of male elders and other similarly vocal elites, great care - involving methodological complexities beyond the scope of the present discussion - must be exercised to canvas the competing viewpoints of less vocal and less advantaged elements in society (Richards & Sharpe 1981).

The present analysis of Sierra Leone rice farming has so far concentrated solely on the group of farmers that finds upland cultivation retains its advantages. Most male heads of household in the southern Sierra Leone case study fall into this category (Johnny 1979). In the northern Sierra Leone study (Karimu & Richards 1980), the sample covered a broader range of categories of wealth and political influence, and included both men and women farmers. Given the choice of upland farming to the exclusion of swamp cultivation, or vice versa, 79% of male farmers would opt for swamp (table 4). (The question is less relevant for women farmers, as very few are in a position to acquire upland in their own right, or to organize the necessary work parties.) As noted above, this was, in most cases, a preference for local swamp management practices, although perhaps 10-15% of all potential recruits to IADP-North had adopted the project's improved swamp package. Farmers were asked to explain their preferences for swamp or upland cultivation (table 5).

Reasons for preferring swamp cultivation included the following points. Yields were frequently described as 'better', though sometimes a rider was added to the effect that this meant 'more reliable' rather than necessarily more rice. (This was an important aspect for woman farmers growing rice to provide a small but steady cash income – male farmers farming to meet family food requirements in the first instance were more likely to stress the 'reliability' of upland intercropping.) Farmers with 'all-year' swamp reported the advantage of double-cropping. (Other farmers may use a swamp which dries out during the dry season for a catch crop of groundnuts or sweet potatoes. This compensates for the reduced potential for intercropping in swamp rice farms.) Swamp rice requires less weeding and pest control, but more labour in mid-season if transplanting is practised. Surprisingly (in view of the fact that, overall, swamp cultivation is more labour-intensive than upland work), most farmers preferring swamp rice claimed the work involved was 'easier'. Easier, however, meant different things to different people. Wealthier, part-time farmers, whose main interests lay in trading or other off-farm employment (civil service, teaching etc.), meant that the work was easier because it was less constrained in terms of timing. Swamp operations are less dependent on weather events. Upland burning and planting are governed by the unpredictable imperatives of late dry-season rainstorms. The timing of swamp operations can be varied much more to fit in with off-farm activities. A further aspects is that, as several informants put it, "it is not necessary to go to the farm every day", partly because of this less-tight scheduling of swamp operations, partly because there is less need, on swamps, for the vigilance-demanding work of pest control, and partly for the apparently paradoxical reason that seasonal labour bottlenecks are so much more marked on swamp than on upland farms (figure 1(a)-(d)). This means that few, if any, farmers can rely on their own labour, and recourse to outside help is almost inevitable. As already noted, hired labour is expensive; but farmers with an off-farm cash income are sometimes in a position to afford it. Thus, although labour input per hectare is increased, the farmer's personal involvement reduces; hence, the work is easier.

Women farmers, and poorer male farmers (a number of whom also prefer the swamp to the upland option) interpret 'easier' rather differently. For them, a major problem is negotiating, each year, for the land and labour necessary for upland shifting cultivation. (Time, negotiating experience, and social prominence all help in securing agreement to use an appropriate piece of land.) Timeliness is of the essence in the upland farm, and work parties are especially critical in ensuring re-

sponse to climatic opportunities.  Women and part-time male
farmers find it difficult or impossible to belong to a work 'company'.  Even for members, the issue of who gets the most
timely assistance is a moot point.  Again 'generosity' and social
status play their part.  In consequence, a less-influential farmer may find considerable strain in membership, and receive
less effective assistance as a result.  A labour company will
fine members who provide the group with substandard food,
and discipline those attempting to evade unfulfilled obligations.
The young, poorer, less-than physically fully-fit farmer, with
unproven ability to match up to the requirements of an 'elite'
company, has the option of seeking out others in a similar
position and setting up a rival group.  Inevitably, such groups
tend to be less efficient, and as a result attract less of the
'outside' cash-earning contracts which provide the better companies with a social security fund and trading capital.  People
in this position gravitate towards swamp farming because it is,
in effect, permanent cultivation – the question of where to
farm each year is solved, and the risk is avoided of being
crowded out by more influential folk should upland begin to be
in short supply – and because less than front-rank companies,
with members distracted by poverty and/or domestic chores,
are better able to cope with the less tightly scheduled operations of swamp farming, even if in consequence returns to
labour are less good than for upland.  This, then, is what
women farmers and poorer male farmers mean when they say
swamp work is 'easier'.  It is more manageable within their
constraints.

By contrast, the typical upland farmer is a sturdy, self-
reliant 'middle peasant', priding himself on his ability to feed
his household and uphold accepted values.  All farmers preferring the upland option cited intercropping possibilities as a
basis for their preference (table 5).  It is possible now to see
that an ideological as well as a nutritional issue is involved.
'Hospitality' – including plentiful and varied food – is an
important aspect of established social values.  Farmers who use
'hospitality' to help construct and maintain the kinds of work
parties needed for efficient management of upland rice farms
are also using the 'varied diet' argument to give further legi-
timacy to social values which have, so far, served them well.
The male, socially influential, full-time farmer who describes
upland farming as 'best' does so because it supports the kind
of social arrangements he is best able to control.  Wealthy,
part-time farmers, with economic and political links to the wider
national system, may be ambivalent about the social values coded
into a system of farm work which they cannot so readily influence.  The 'modernity' of swamp cultivation is thus a useful

ideological lever in attempting to gain greater control over social relations of production, even if, without an abundant supply of cheap hired labour, the farming system itself poses certain practical difficulties. It is likely, however, that off-farm sources of income are great enough to solve this problem through paying for 'surplus' company labour. Extensive networks of rural patron-client relationships underpinning local politics ensure special deals for powerful patrons. If a company leader 'likes' a local 'big man', favourable terms will be arranged.

Figures 1(a)-(d) return us to the quantitative approach. The median-sized household in the IADP-North project area has about nine members, produces approximately 2500 kg milled rice per year, and is capable of generating, in normal circumstances, 60 workdays worth of labour input per month (Karimu & Richards 1981). Figures 1(a)-(d) show the monthly distribution of labour required to produce 2500 kg of rice on a typical farm, using upland, local swamp, and improved swamp techniques. (For details concerning data and estimation procedures see Karimu & Richard 1981.) Labour bottle-necks - labour requirements beyond the farmer's capacity - occur under all three farming systems. Harvest labour peaks are not serious, because when food is forthcoming so is labour. The transplanting bottleneck on 'improved' swamp farms is much more serious, since it coincides with the period of pre-harvest 'hunger'. Poor farmers go into debt to purchase food and cannot afford hired labour as well. Wealthier, part-time farmers can hire labour. It is important to note, however, that local swamp farming is perhaps better than improved swamp for the farmer paying for labour out of dry season off-farm income, since not only is the peak requirement smaller, but the total length of cultivation season is shorter, thus maximizing scope for non-farm activities. The full-time farmer wishes to minimize under-employment in the dry season and labour peaks later in the year. The best way to do this is to combine upland and local swamp cultivation in a ratio of 2:1 (figure 1(d)), which, perhaps unsurprisingly, is what the bulk of 'middle peasants' in the IADP-North project appear to do.

## 3. SUMMARY AND CONCLUSION

In summary, this paper has shown that measurements of agricultural work and what people say in evaluating that work should not be taken in isolation from an analysis of the group interests involved in agricultural practice and agrarian change. 'Middle peasants' continue upland rice farming because this

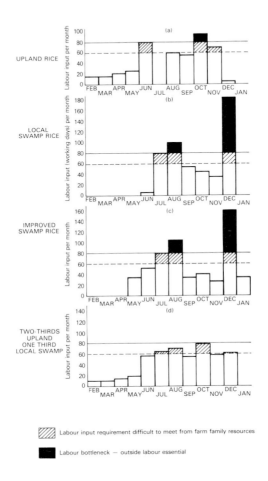

FIGURE 1(a)-(d). Seasonal labour input profiles – various Sierra Leone rice-farming systems.

helps to reinforce a set of social values they are well-placed to capitalize upon. Swamp farming better reinforces the interests of wealthier, part-time farmers with significant off-farm business interests, and perhaps political connections as well. The interests of this group coincides with the government's concern to increase rice supplies to urban areas; hence the emphasis on swamp farming in recent rural development programmes. Women and poor male farmers are attracted to swamp cultivation as an answer to the social disadvantage they suffer in respect to the organization of upland farming, but are unable to transcend a new set of contradictions built into 'improved' swamp cultivation. They are better placed using local swamp techniques. These distinctions should not be seen as hard and fast. A forced choice between upland and swamp is an unfair question, since most farmers get by using a combination of both farming systems. The unfairness, or inappropriateness, of this question is in effect a further example of a failure to dig out ideological presuppositions embedded within the researcher's own frame of reference. The recognition that these presuppositions themselves reflect 'group interest' suggests the need for the 'scientific' study of 'work' to move beyond positivism and to develop greater dialogue with groups studied concerning the proper objectives of investigation.

## 4.  ACKNOWLEDGEMENTS

The data in this paper derive from a joint study carried out by Michael Johnny, Braima Josiah, John Karimu and Paul Richards. The study was funded by the Inter-University Council for Higher Education Overseas, London.

## REFERENCES

Dunsmore, J.R., Blair Rains, A., Lowe, G.D.N., Moffat, D.J., Anderson, I.P. & Williams, J.B., 1976, The Agricultural Development of the Gambia: an Agricultural, Environmental and Socio-economic Analysis, Land Resource Study 22 (Surbiton, Surrey: Land Resource Division, Ministry of Overseas Development.)

Habermas, J., 1972, Knowledge and Human Interests (London: Heinemann).

Johnny, M.M.P., 1979, Traditional Farmers' Perceptions of Farming and Farming Problems in the Moyamba Area (Sierra Leone (MA Thesis, University of Sierra Leone).

Johnny, M.M.P., Karimu, J. & Richards, P., 1981, Swamp and
    upland rice farming systems in Sierra Leone: the social
    context of technological change, Africa, 51, 596–620.
Johnny, M.M.P., & Richards, P., 1981, Playing with facts: the
    articulation of "alternative" viewpoints in African rural
    development. In Folk Media and Development, edited by
    R. Kidd and N. Coletta.
Karimu, J.A. & Richards, P., 1981, The Northern Area
    Integrated Agricultural Development Project: the social and
    economic impact of planning for rural change in northern
    Sierra Leone, Geography Department, School of Oriental
    and African Studies, Occasional Papers, 4.
Richards, P. & Sharpe, B.J., 1981, The Social Construction of
    Labour: Two West African Case Studies (University College,
    London: Anthropology Department), unpublished.
Sierra Leone Central Statistics Office, 1971, Agricultural
    Statistical Survey of Sierra Leone 1970, 1971. (Freetown:
    Government Printer).
Spencer, D., 1975, The economics of rice production in Sierra
    Leone: i. upland rice. Bulletin, 1. (University of Sierra
    Leone, Njala University College: Department of Agricultural
    Economics and Extension).

# ENERGETICS AND ECOLOGICAL
# ANTHROPOLOGY:   SOME ISSUES

P. Burnham
Department of Anthropology,
University College
Gower Street, London WC1E 6BT

## 1.   INTRODUCTION

The idea that human society can be understood as a system
constructed by human effort and that measurement of this effort can provide a means of studying human society has a long
intellectual pedigree.    Yet in recent years, under the stimulus
of the energy crisis and ecological politics, the study of energy
use in human society has taken on renewed vigour.    The biologist H.T. Odum's Environment, Power and Society (1971) provides an extreme and much-discussed attempt to understand the
entire world system in energy terms.    By conflating the concepts of work, power and energy, and by viewing political and
religious institutions as "master controls of human systems"
(p.205), Odum (1971) attempts to formulate an analytical language which will permit him to discuss any aspect of human history in terms of energy relations.    Apparently undaunted by
the pitfalls of reductionism, Odum (1971) makes his position
clear:   "Although nearly everyone is familiar with power ratings
of household appliances and automobiles, our educational system
has rarely emphasized that the affairs of man also have quantitative power ratings and that the important issues of man's
existence and survival are as fully regulated by the laws of
energetics as are the machines" (p.26).    Predictably, Odum's
handling of social phenomena such as economics, politics and
religion proves to be breathtakingly naive, combining an energy-
based economism with a thinly disguised systems-theory rationale
for the American political system, and adding a messianic call
for "relevant religious dogma" based on a set of "Ten Command-
ments of the Energy Ethic" (p.244).    Although acclaimed by
certain anthropologists as "particularly useful for human ecological studies" (Morren 1977, p.274), it is my view that Odum's
"energy language" obscures rather than clarifies the social processes at work in human ecological systems.    Indeed, this is
a criticism that can be levelled at the overwhelming majority of

studies of human societies from the energetics perspective, a
point that I shall document repeatedly in the present paper.
    Mention of the socially defined nature of labour in human
society inevitably brings to mind the work of Marx and especially his labour-time theory of value.   Although not the inven-
tor of this general idea, Marx is certainly its best known ex-
ponent, and at the risk of stating the obvious, I believe it is
important to emphasize that: (a)  Marx's concept of socially necessary labour time is different from the ecologist's notion of
energy;   (b)  it is an abstract concept which is not even equivalent to the measurable work carried out by an individual in a
particular situation;   and (c)  Marx himself only expected that
value would be calculable on the basis of socially necessary
labour under very specific economic conditions in certain historical periods.   Despite these considerations, the notion of a
labour-time basis for value continues to exercise a fascination
for certain anthropologists and has resulted in several field
studies in which such calculations have been attempted (e.g.,
Salisbury 1962, Godelier 1972, Cook 1970).   The results, while
uneven in quality and far from convincing, at least have the
merit of illustrating Marx's arguments about the complex social
determination of value and the lack of any necessary relation
between the physical effort expended in the production of a
good and its exchange value.   Such studies also underline the
simplistic character of assertions such as Odum's (1971) that
the "value (of a good) is the time integral of the flow of potential energy expended in work" (p.185).
    Turning to the main stream of development within ecological
anthropology, a focus on the concept of energy via the study
of societies as systems of energy flow has always been a characteristic of ecological anthropology, and was at the heart of
the developments that led to the founding of this sub-discipline.
The work of Leslie White was particularly seminal in this regard.
In White's (1949) formulation, cultural evolution is determined by
the control of a progressively greater amount of energy, which
is utilized by human societies via improvements in technology.
As a result, according to White, societies could be graded in
relation to their progress in energy control, a scale which was
directly comparable to the concept of evolutionary grades in
biological evolution (see Sahlins & Service 1960).   Social factors were definitely of secondary importance in this scheme.
"Technology is the hero of our piece", wrote White (1949, p.390).
"Thus we see that the social system of a people is at bottom
determined by the use of the technological means of subsistence
and of offense and defense.   Those social institutions not directly related to the technology are related indirectly;   they serve
to coordinate the various sectors of society with one another and

to integrate them into a coherent whole" (pp.376,377).

White's writings on the significance of technological control of energy as the determining factor in cultural development were largely restricted to the general theoretical level, and it has been left to his followers to attempt to apply his perspective to particular ethnographic cases.    In this study of what Sahlins (1960) termed "specific evolution", one of the main problems to be addressed was as follows – if energy capture via improved technology is the motor generating evolutionary progress in society, how is this motor linked to the rest of the social system?    The answer, according to Marvin Harris, is that societies have a propensity to maximize the usage of available energy, developing their patterns of social activity rationally so as to fulfill social needs.    This, in Harris's view, is the character of the adaption process in human society, and it justifies proceeding from "the assumption that all of the energy which a primitive group produces above its thermodynamic subsistence level is probably a biosocial necessity" (Harris 1959, p.193).    Declaring himself against the idea that societies produce "superfluous" surpluses, he goes on to argue that, "If the notion of superfluous products of any sort is at odds with the theory of functionalism, the notion of a superfluous production of food is at odds both with the theory of functionalism and the theory of evolution" (p.192).

One of the results of the theoretical position adopted by White, Harris and others has been to generate what Brookfield (1972) has termed a "calorific obsession" in empirical studies carried out within the cultural materialist and ecological anthropology tradition (p.46).    The measurement of food production, especially the measurement of starch staple food production, has been viewed as the analytic key which will unlock the rationality of the social systems under study.    Of almost equal interest, for the same reason, has been the measurement of available protein supplies.    We shall be considering the methodological and theoretical implications of this position later in this paper.

But of more fundamental theoretical importance has been the willingness of authors like Marvin Harris to continue to espouse a naive functionalism which has repeatedly been discredited (see for example, Burnham 1973, Friedman 1974).    As Friedman (1974) notes, "The new functionalism is fundamentally the same as the old functionalism except that the field of application has changed, the interest now being to show the rationality of institutions with respect to their environments rather than to other elements in the society, (p.457).

In contrast to Harris's attempt to develop White's theory by positing a continual tension towards expansion via maximization

of energy production and utilization, Vayda, Rappaport and
their students have preferred to focus on the tendencies towards
equilibrium in the New Guinea social systems they have studied.
In Rappaport's now-famous study of the Maring people, Pigs for
the Ancestors (1968), ritual and warfare are considered to function as regulators of the process of extracting energy from the
environment, ensuring that the carrying capacity of the locality
is not exceeded.    In particular, the size of the Maring pig herd
and the amount of agricultural production needed to sustain these
animals are said to be regulated via a complex ritual and warfare
cycle, in which large numbers of pigs are periodically sacrificed
in major festivals.

Rappaport's Maring study has stimulated more than a decade
of theoretical debate (see, for example, Friedman 1974, 1979,
McArthur 1974, Burnham 1979) and although Rappaport (1977)
continues to defend his position, going as far as to define the
discipline of ecological anthropology as the study of self-persist-
ing systems, there is an increasingly wide recognition elsewhere
in the discipline that this single-minded focus on equilibrium is
mistaken (Vayda & McCay 1975).    Further discussion of these
theoretical issues would take us far from our assigned subject,
and I propose now to turn to the problems of method encountered
in Rappaport's and other studies of energetics in human
social systems.


## 2.    METHODS OF STUDY OF ENERGETICS IN SOCIAL SYSTEMS

General theoretical considerations aside, a major rationale for
studying energetics in ecological anthropology has been one of
method.    Bayliss-Smith (1977) has argued, for example, that
"despite the various criticisms of an energy approach, energy
does have the advantage of being a fundamental and value-free
measure of quantity" (p. 318).    Morren (1977), writing in the
same volume, echoed Bayliss-Smith's words when he wrote, "An
energetic approach to environmental problems and human responses to them is advantageous because it makes possible a high
degree of intersubjectivity and quantification" (p.277).    But
behind these apparently uncontentious statements lurks a host
of methodological and conceptual problems which may fundamentally compromise the feasibility and utility of studies of energetics in ecological anthropology.

Thanks to several pioneering studies of energy systems in
human societies, especially that of Rappaport just mentioned and
R. Brooke Thomas's work as a part of the Man in the Andes pro-

ject (Baker & Little 1976), we have gained considerable understanding of the methodological problems associated with such
studies.    To begin with, quantification of energy flows in even
a relatively non-complex human society has proved to be excep-
tionally time-consuming, especially in view of the necessity for
adequate sampling.    In all likelihood, the sufficiently detailed
measurement of such a real-life system is beyond the ability of
a single anthropologist working for even an extended period in
the field.    Then, too, there is the problem of the multidisciplinary competences required of a single researcher engaged in
human ecological field study - a problem which Richard Feachem
(1977) has aptly summed up in the title of his recent article
"The Human Ecologist as Superman".    Margaret McArthur (1977),
commenting on Rappaport's work among the Maring, has drawn
attention to the same issues.    Moreover, she has listed numerous points of error or weakness in Rappaport's nutritional data
and interpretations which result from his lack of specialist training in human nutrition and from the inherent difficulties of field
studies on this topic.    Although McArthur does not consider in
any detail the horticultural aspect of Rappaport's study,similar
critiques are possible here, due to the technical difficulties encountered in assessing swidden yields, land-use patterns, and
other relevant variables.    (See Brush (1975) for a fuller discussion on these points).    Indeed, Rappaport himself (1977)
has made reference to such problems in his own data in a recent
defence of the theoretical approach he used in Pigs for the Ancestors.
     Faced with such problems, it is only natural that workers in
the field of human energetics should increasingly turn to interdisciplinary teamwork to carry out field research.    Such efforts
were facilitated by the IBP programme, and of those field studies,
the Nunoa project in the Peruvian Altiplano is generally considered
to have achieved the most satisfactory overall results in the field
of human energetics (Vayda & McCay 1975, Morren 1977, Harrison
1979).    Having access to substantial numbers of biologically
trained specialists, a team of field assistants, and adequate local
laboratory facilities, the Andean project was naturally able to
overcome some of the problems which had plagued Rappaport's
one-man study.    However, the logistic and financial requirements
for a complex project of this nature make it very unlikely that
such large-scale efforts will frequently be repeated.    In view of
this, and particularly in the context of increasingly tight research
budgets, it seems doubly important to assess carefully the theoretical and methodological 'pay-offs' of a major project of this
type.    Ecological anthropologists seem only too prone to think
that if they only were able to measure adequately a human energy-

-flow system, a major advance would automatically follow in
general theoretical and conceptual terms.

Leaving the theoretical issues until later, we must first consider the methodological problems that affected a project even
of the size of the Andean one.    One of these was that, in practice, it proved to be very difficult to assess accurately the energy requirements for the population's nutrition.    Although the
difficulties of local travelling made adequate sampling on an annual basis impossible, and the limited time available for the project ruled out assessment of year-to-year variations, these were
not really the most fundamental problems.    A major problem
was that Thomas's energy-accounting methodology required an
accurate assessment of a notional minimal level of caloric need,
and fixing a figure within reasonably narrow confidence limits
proved a thorny issue.    Other nutritionally qualified participants in the present conference, such as Rivers (p. 85) and
Ferro-Luzzi (p. 115), have already commented on this point generally, as has McArthur (1977).    Basing my views on such
expert opinion, it is my understanding that it is quite difficult
to establish a figure for a minimally adequate energy intake for
a population on the basis of field studies, and that a whole host
of factors and corrections reflecting such variables as sex, age,
body weight, temperature, activity patterns, lactation, roughage
in the diet, parasitic infections and other diseases, as well as
possible physiological adaptations to particular dietary regimes,
need to be taken into account.    Certainly, it seems widely accepted that the FAO/WHO 1973 standards are set at a level which
considerably exceeds such a notional nutritional minimum and,
in any case, these standards were never meant to be used in this
way.

In attempting to assess whether or not energy was a limiting
factor for the Nunoa population, therefore, Thomas had to decide
whether to accept the results of his own nutritional survey, however imperfect, as reflecting the minimal caloric needs of that
population, or whether to adjust these figures in some way, taking into account the somewhat conflicting findings of other nutritional and health surveys of the same region (Gursky 1969,
Picón-Reátegui 1976, Thomas 1976).    In making his assessment,
it is not clear to what extent Thomas took into account the patterns of social stratification in the region nor how he related the
question of individual caloric balance to the social functioning of
the Nunoa population as a whole.    I am not suggesting that,
for those wishing to work within a theoretical framework based
on energetics, the answers to such questions are easy ones.    In
fact, I am suggesting just the opposite;    namely, that a major
margin of uncertainty concerning energy requirements, and

indeed energy production, will inevitably plague such projects. But these are only a few facets of the problem, as we will see when we look at the Andean project further.

In any study of energetics in human social systems, one is forced to make numerous simplifying assumptions and, in particular, to draw an artificial boundary around the system under study.   In principle, then, one can determine the energy available within the system by calculating local energy production and adding and subtracting energy passing across the boundaries of the system.   While this is fine in theory, such a procedure is rarely carried out satisfactorily in practice, and in this regard, the Andean project is open to criticism.

The social system of the Andean community under study is a complex and highly stratified one, with a dominant feature of the local society being the 104 haciendas, or family estates, owned by both resident and absentee landlords and worked by Quechua serfs or sharecroppers.   Although geographically isolated, Nunoa is nonetheless linked to an export economy based mainly on wood and other herd products (Escobar 1976).   Locally, herding is the main form of production, and a number of the settlements in the study area were located at altitudes at which cultivation was impossible due to frost.   In fact, less than 2% of the land in the district of Nunoa is suitable for agriculture (Little & Baker 1976).   Nonetheless, in all communities studied by the Andean project team, vegetable food made up the great majority of the daily diet, and often a sizeable amount of this food was imported from outside the region under study (Picón-Reátequi 1976).   Further relevant facts could be listed, but the set just mentioned are sufficient to make the point that an accurate calculation of energy flows in such a system would be extremely difficult, even under the best of research conditions. In the event, as Thomas (1976) admits, most of the data relevant to the above-mentioned features of the system are not included in his model:   "In this report, socioeconomic units larger than the nuclear family have been neglected.   Moreover, the hacienda, which gets access to about an additional 1,000,000 kcal produced by the family, has not been discussed.   Such production costs the family over 40,000 kcal" (p.403). Neither of these values appear in Thomas's table of mean caloric consumption estimates for Nunoa.

Not only does Thomas neglect the exploitation by the hacienda owners of their serfs, with its attendant implication that this may be one of the primary limiting factors on Indian access to energy, but he also fails adequately to address the problem of inter-familial, inter-zonal, inter-regional, and international exchange as it affects Indian access to energy.   If Indians are mainly producing herd products and are mainly consuming vegetable

foods, substantial exchange is occurring both within and across
the boundaries of the system.    What factors determine the exchange rates for, say, wool versus wheat flour?    If, as Escobar (1976) notes, "The heyday of the hacienda system in Nunoa
and the rest of the altiplano was between the two world wars
when the price of wool in the international market was very
high" (p.70), could it not be the case that the energy available
to these Andean Indians is determined as much by fashion in
London and New York as it is by the local environmental stresses
in Nunoa?    International trade aside, there is also a substantial
inter-zonal and local 'meat for potatoes' trade which is equally
governed by an exchange rate that Thomas fails to analyse.    It
is worth emphasizing that such exchange rates, when calculated,
can prove to be highly variable over time.    Drawing an example
from my own fieldwork, over a period of five years in one locality
in Cameroon where I have carried out research, the exchange rate
between cattle and manioc, the starch staple, varied by as much
as 100% (Burnham 1980).

Here, then, is another major source of indeterminacy in Thomas'
attempt at an energy-flow analysis of an Andean community.    One
could continue to accumulate such inaccuracies and margins of
error, but I believe enough has been said to illustrate my point.
Even in a well endowed and interdisciplinary project which runs
over a substantial period of time, the goal of producing a sufficiently accurate accounting of energy flows within a social system to permit the formulation of meaningful statements about energy
efficiencies and the limiting role of energy supply remains unrea-
lizable.    But more fundamental than this, as I have already in-
timated and now will go on to argue, even if we did dispose of
such a comprehensive and accurate portrayal of a human society
in energy terms, we lack a convincing theoretical framework which
would allow us to interpret these data.

For at least several decades now, the dominant explanatory
paradigm in ecological anthropology has been the functional/adapt-
ational one borrowed from biology.    In this framework, social
factors are interpreted in relation to their contribution to the
successful functioning of the total system.

As I have already explained at the outset of the paper, there
are at least two versions of this theory of social adaptation in use
in anthropology.    In Marvin Harris's version, the functioning of
the social elements in the system is thought to be usually positive,
due to an alleged propensity for human societies to maximize usage
of available resources in a rational manner.    In Rappaport's version, the functions of the social elements in the system are con-
ceived of as being potentially positive or negative and are assessed
by the neo-functionalist technique of comparing observed data value

with independently established tolerance limits.    In practice,
however, due to Rappaport's espousal of the concept of control
hierarchy (cf. Friedman 1979), there is also a tendency in this
version of adaptation theory to stress the positive functional
contribution of the elements of a system.

As I indicated earlier, this theoretical framework and the related ethnographic literature have been extensively commented
upon and criticized, with particular attention being placed on
the inadequacy of the concept of adaptation as applied to social
behaviour.    In the present context, it may be relevant to demonstrate the problems associated with the adaptation concept
as used in Thomas's Andean energy-flow study, for although the
conceptual problems in this study are little different from those
already noted in other studies in ecological anthropology, and
one therefore risks stating the obvious, Thomas's project continues to be cited as a positive model for this type of work
(Vayda & McCay 1975, Morren 1977, Harrison 1979).

If you will pardon a lengthy quotation, according to Thomas
(1976), "The concept of 'adaptation to the (energy) flow system'
suggests that hypocaloric stress is effectively counteracted and
that energy produced is being utilized in an efficient manner.
It is therefore possible to regard adaptation under limiting conditions as a strategy that attempts to minimize the risk of serious
disruption in flow and to maximize long-term energetic efficiency
(energy production/energy expenditure).    The assessment of
specific adjustments to energy flow may be approached in the
same way.    If alternative responses or strategies are available
to a group, those with the highest long-term energetic efficiency
may be regarded as the most adaptive, since higher energetic
efficiency generally makes serious hypocaloric stress less likely.
Such a definition does not imply causation, simply a functional
relationship".

Long-term energetic efficiency is therefore the criterion of
"adaptation" in Thomas's study, and we are told, for example,
that the herding of animals by children rather than adults is
more energy-efficient and that, therefore, "high fertility appears
adaptive in terms of the family".    Of course, as Thomas also
recognizes, demographic pressure is a threat to the Nunoa system,
which is relieved under present circumstances by high mortality,
especially in the early age groups, and by emigration.    And so,
if increased fertility generates both a larger supply of child
labour and a higher mortality rate (to mention only several of the
linked variables in the Nunoa system), we are left with the in-
tractable problem of assessing the relative importance of these
effects for the net selective advantage of large family size. Thomas
himself has attempted to sidestep this problem by isolating the
high fertility—child labour—energy efficiency linkage from the

wider social context in which it exists, a solution which flies
in the face of the very concept of 'system' which he has otherwise taken to be central for the rest of his study.    Judged
solely in relation to energy efficiency, use of child labour and
large family size are said to be the most 'adaptive' strategies
available to a family, but Thomas's definition of adaptation bears
little, if any, relation to the reproductive fitness-based defini-
t'ᵤn of the biologists.    Particularly in interdisciplinary research
such as the Nunoa project, in which biologists and social scientists are collaborating, I consider it misleading and dangerous to
allow a central concept such as adaptation to be used in different
ways simultaneously.

On the other hand, Thomas's use of the phrase "adaptive
strategies" may be meant to imply that Nunoa couples consciously
make decisions about family size in relation to the consideration
of energy efficiency.    Yet if Thomas really meant to focus on
the rationality of Nunoa family-planning practices in relation to
energy expenditure, it is difficult to see what is gained analytically by substituting the term "adaptation" for the term "rationality" or the like.    Moreover, since it seems quite improbable
that such decisions are taken with this factor solely in mind,
Thomas's energy-based rationality is cognitively unrealistic.

Seen in a comprehensive framework, Thomas's energy-efficiency
based definition of adaptation encounters other difficulties.    The
finding that pastoral economies make plentiful use of child labour
for herding activities is hardly a novel one;    indeed, it is a
commonplace in many of the studies of African pastoralists with
which I am familiar.    In some of these societies, energy might
be said to be a limiting factor, in others it definitely is not.  Can
the use of child labour always be said to be an energy-efficient
adaptation?    Is it an adaptation even when energy is not limiting?
And why do certain pastoral communities not make use of child
labour in this way?    Thomas's concept of adaptation, narrowly
defined in terms of energetic efficiency, cannot really hope to
deal with these questions, and in the absence of an adequate
general concept, description of Nunoa social behaviour as "adaptive" is nothing more than a sleight-of-pen, conveying the impression of a unified theoretical framework for biological and
social ecology where none in fact exists.

By way of conclusion, then, I feel that I should perhaps
apologize for what to some will have seemed a very nihilistic
paper.    But I believe it is necessary to clear the decks of the
energy-based grand theories which promise to solve our problems
using a single variable.    Such theories have been a major
stumbling block for the development of a truly social ecology.
Here and there over the last decade in the literature of ecological anthropology, that is to say among the proponents of energy

theories themselves, one can begin to detect a movement of opinion in this direction.     Thus, in a recent review of ecological anthropology, Vayda & McCay (1975) accept that a "calorific obsession" has been one of the field's main characteristics and limitations.     But the conclusion that Vayda & McCay draw from this realization is certainly not the one that I would recommend.     In effect, their proposal is that ecological anthropologists should merely trim their sails in using an energy systems approach and restrict such analyses to cases where energy is found to act as a limiting factor.     According to Vayda & McCay (1975), "In the case of people for whom energy and its translation into food and fuel calories do appear to be major limiting factors, energy flow studies can be expected to contribute significantly to our understanding of how the existential game is played" (p.296).     In fact, I would argue that few studies in the field of anthropology have contributed less to our understanding of the "existential game" than have energy-flow studies, based as they are on such unrealistic assumptions about human behaviour.     The Second Law of Thermodynamics notwithstanding, i fear that we shall make little headway in developing an understanding of the implications of energy flow in human societies until ecological anthropology takes seriously the proposition that energy can never be treated as a limiting factor which operates independently of the social context.

## REFERENCES

Baker, P.T. & Little, M.A., 1976, Man in the Andes (Stroudsburg, Pennsylvania:   Dowden, Hutchinson & Ross, Inc.).

Bayliss-Smith, T.P., 1977, Energy use and economic development in Pacific communities.     In Subsistence and Survival, edited by T.P. Bayliss-Smith and R.G. Feachem (London: Academic Press), pp.317-359.

Brookfield, H.C. 1972, Intensification and disintensification in Pacific agriculture:  a theoretical approach.     Pacific Viewpoint, 13, 30-48.

Brush, S., 1975, The concept of carrying capacity for systems of shifting cultivation.     American Anthropologist 77 (4), 799-811.

Burnham, P.C., 1973, The explanatory value of the concept of adaptation in studies of culture change.     In The Explanation of Culture Change, edited by C. Renfrew (London:  Gerald Duckworth), pp.93-102.

Burnham, P.C., 1979, Permissive ecology and structural conservatism in Gbaya society.   In Social and Ecological Systems, edited by P.C. Burnham and R.F. Ellen (London:  Academic Press), pp.185-202.

Burnham, P.C., 1980, Changing agricultural and pastoral ecologies in the West African savanna region.   In Human Ecology in Savanna Environments, edited by D.R. Harris (London: Academic Press), pp.147-170.

Cook, S., 1970, Price and output variability in a peasant-artisan stoneworking industry in Oaxaca, Mexico.   American Anthropologist, 72, 776-801.

Escobar M., G., 1976, Social and political structure of Nunoa in Man in the Andes, edited by P.T. Baker and M.A. Little (Stroudsburg, Pennsylvania: Dowden, Hutchinson & Ross, Inc.), pp. 60-84.

Feachem, R.G., 1977, The human ecologist as Superman.   In Subsistence and Survival, edited by T.P. Bayliss-Smith and R.G. Feachem (London: Academic Press), pp.3-10.

Friedman, J., 1974, Marxism, structuralism and vulgar materialism. Man (N.S.), 9, 444-469.

Friedman, J., 1979, Hegelian ecology: between Rousseau and the World Spirit.   In Social and Ecological Systems, edited by R.F. Ellen (London: Academic Press), pp.253-270.

Godelier, M., 1972, "Monnaie de sel" et circulation des marchandises chez les Baruya de Nouvelle-Guinee.   In Horizon, Trajets Marxistes en Anthropologie, edited by M. Godelier (Paris: Maspero), pp.259-293.

Gursky, M., 1969, A dietary survey of three Peruvian highland communities (Pennsylvania State Univ., University Park, Pennsylvania: M.A. Thesis).

Harris, M., 1959, The economy has no surplus?, American Anthropologist 61, 185-199.

Harrison, G.A., 1979, Views from three other disciplines: biological anthropology.   In Social Anthropology of Work, edited by S. Wallman (London: Academic Press), pp.37-42.

Little, M.A. & Baker, P.T., 1976, Environmental adaptations and perspectives.   In Man in the Andes, edited by P.T. Baker and M.A. Little (Stroudsburg, Pennsylvania: Dowden, Hutchinson & Ross, Inc.), pp.405-428.

McArthur, M., 1974, Pigs for the ancestors: a review article, Oceania 45, 87-123.

McArthur, M., 1977, Nutritional research in Melanesia: a second look at the Tsembaga. In Subsistence and Survival, edited by T.P. Bayliss-Smith and R.G. Feachem (London: Academic Press), pp.91-128.

Morren, G., 1977, From hunting to herding: pigs and the control of energy in montane New Guinea. In Subsistence and Survival, edited by T.P. Bayliss-Smith and R.G. Feachem (London: Academic Press), pp.273-315.

Odum, H.T., 1971, Environment, Power and Society (New York: John Wiley).

Picón-Reátequi, E., 1976, Nutrition. In Man in the Andes, edited by P.T. Baker and M.A. Little (Stroudsburg, Pennsylvania: Dowden, Hutchinson & Ross, Inc.), pp.208-236.

Rappaport, R., 1968, Pigs for the Ancestors (New Haven: Yale University Press).

Rappaport, R., 1977, Ecology, adaptation and the ills of functionalism, Michigan Discussions in Anthropology, 2, 138-190.

Sahlins, M., 1960, Evolution: specific and general. In Evolution and Culture, edited by M. Sahlins and E. Service (Ann Arbor: University of Michigan Press), pp.12-44.

Sahlins, M & Service, E., 1960, Evolution and Culture (Ann Arbor: University of Michigan Press).

Salisbury, R.F., 1962, From Stone to Steel (Melbourne: University of Melbourne Press).

Thomas, R.B., 1976, Energy flow at high altitude. In Man in the Andes, edited by P.T. Baker and M.A. Little (Stroudsburg, Pennsylvania: Dowden, Hutchinson & Ross, Inc.), pp.379-404.

Vayda, A.P. & McCay, B., 1975, New directions in ecology and ecological anthropology, Annual Review of Anthropology, 4, 293-306.

White, L., 1949, The Science of Culture (New York: Farrar, Straus & Cudahy Co.).

# THE USE OF MODELS IN ANTICIPATING EFFECTS OF CHANGE ON HUMAN POPULATIONS

R. Brooke Thomas[1], S.D. McRae[2] and P.T. Baker[3]

[1]Department of Anthropology, University of
Massachusetts, Amherst
[2]Department of Earth Sciences, University of California,
Riverside
[3]Department of Anthropology, Pennsylvania State University,
University Park

## 1. ANTICIPATING CHANGE

As change in human-environmental systems proceeds at an unprecedented rate and scale, and frequently has unanticipated and disruptive effects on human wellbeing, social organization and environmental quality, scientists and planners alike are confronted with an urgent need to uncover regularities about this process.

In attempting to understand processes of change, it has become abundantly clear that our traditional analytical approaches are overly restrictive. The information base needed extends far beyond any one area of expertise. Conditions we are asked to assess have no historical precedents to guide our expectations. Information, once collected, is outdated by the time of publication. Experimentation with human-environmental systems is frequently unfeasible or unethical. In short, information we are able to bring to understanding change has often lacked realism or relevance because it does not adequately take into account the many variables interacting in dynamic systems.

If solutions can indeed be found they will have to rely upon analytical procedures which can incorporate a broader information base, drawn from human biology, social science and environmental science; emphasize interconnectivity between multiple variables operating in complex systems; and yet reduce the complexity to a consideration of critical variables, in order that the problem becomes analytically tractable. Since our capabilities for predicting most forms of change remain rather rudimentary, it seems that analytical techniques which could anticipate alternative consequences of a change might contribute significantly. In this respect, modelling of complex systems serves as one such technique.

## 2. MODELS

Models are partial representations of perceived reality which depend heavily upon a visual component to demonstrate inter- actions between essential features of a system – interactions not easily comprehended by other means.

Thus, mannequins and road maps are types of model representing aspects of reality important to particular objectives. If effective, these and other models represent degrees of general- ity, reality and precision appropriate to their purpose, and ignore irrelevant features. Models are therefore based on a series of assumptions concerning what are important components of a system and how these relate to one another. Since the assumptions and relationships between components reflect the state of the art, models should be regarded simply as heuristic tools whereby complex phenomena can be assessed, and from which testable hypotheses can be generated.

When system variables and rates of flow between them are quantifiable, a model can function in the following manner. (Shantzis and Behrens 1972, p.259).

"(a) The theory of complex feedback loop systems can aid in understanding and organizing the important causal re- lationships in the observed system.

(b) Analysis of the model's sensitivity to changes in its parameters can indicate where precise observations or measure- ments are important and where large observational errors are relatively unimportant in understanding overall societal functions.

(c) The model provides a framework within which one can raise new questions and perceive missing information to design further studies more efficiently.

(d) Analysis of the model can provide information on the behavioural implications of observed relationships outside the range of parameter values historically observed. Thus, it is useful for testing the probable effects on the society of new technologies or social policies."

Simulation modelling can therefore be used as a device to gain insights into a system's operation. In addition, it is a useful technique when, for ethical, financial or time considerations, experimental manipulation and observation of the system are not possible (McRae 1979). Pros and cons of simulation are dis- cussed in detail by DeGreene (1973), Forrester (1968), Holling (1973), Holling & Chambers (1973), Lee (1973), Shannon (1975), Walker (1973) and Watt (1968).

In constructing a model, several considerations, reviewed by Levins (1966), are noteworthy. First, it is difficult to simultaneously emphasize generality, reality and precision. Applied scientists, for instance, generally attempt to emphasize precision and realism, viewing generality as not instructive over the short-term periods they are interested in. Conversely, evolutionary scientists and archaeologists, who compare systems over longer stretches of time, frequently sacrifice precision for generality and realism. Here, qualitative rather than quantitative changes of variables interacting with one another are of greater importance.

Like other analytical techniques, models are constructed to address specific problems. By design, they must leave out a lot in order to concentrate on essential features; hence, they are bound to be inaccurate and incomplete when judged from a perspective different from that from which they were designed. General models, for instance, have three kinds of imprecision (Levins 1966). Variables which have small effects on the system, or large effects which occur only rarely, are not considered. In stressing qualitative properties, they are vague about the exact form of mathematical functions. As variables are collapsed into higher-order or more general variables, this destroys information about lower-order phenomena. The collapsing of various cultigens into a variable labelled 'crops', or of measures of adaptiveness into an energetic efficiency parameter serve as examples. By considering many cultigens as crops, a many-to-one category, we are denied access to information concerning their differences, and choose to emphasize characteristics shared between them.

A final consideration in modelling concerns the accuracy of the model constructed. Various forms of model validation reviewed by DeGreene (1973) follow.

Face validity: does the model seem to be related to the behaviour of the system?

Content validity: are processes (economic, demographic) in the real system reflected in the model?

Content validity: are logical relationships and sufficient data present?

Consensual validity: do people who know the real system feel the model fulfills its purpose?

Concurrent validity: how closely is model behaviour associated with that of the real system at approximately the same time?

Predictive validity: does the model accurately predict the consequences of future changes in the real system?

The last two forms of validation are more rigorous. Predictive validity requires two independent sets of data, since one cannot validate a model from the data with which it was built.

As Levins (1966) notes, a model is neither a hypothesis nor a theory. It is built from relationships between variables represented in reality. What is important, however, is not that such relationships exist, but how important they are in influencing systemic functioning. Validation of a model in an ultimate sense, therefore, depends on its capacity to generate testable hypotheses about the interconnectiveness of multiple variables interacting in complex systems. In this sense, the model is a transitory phenomenon. Once it has served its purpose, it is usually discarded or altered in order to more precisely address a problem.

Since the results generated by models are influenced by simplifying assumptions, the selection of critical variables, and determined relationships between these variables, it is difficult to know the relative effects of these factors. It is only when models designed in different ways yield similar results, that we can refer to this consistency as a general theorem.

## 3. SELECTION OF A CURRENCY

In modelling a system where quantification of variables and rates of change between them is desirable, selection of an appropriate flow currency needs to be considered. In human systems, important flows track monetary exchange, energy, limiting nutrients, materials and information. If realism is to be maximized, as many flows as possible should be monitored. However, there is considerable heuristic value in initially presenting a system in terms of a single currency flow.

Selection of the type of flow depends on the following criteria: (a) the importance of the variable in explaining adaptive phenomena; (b) the ability to measure the variable; (c) the ability of the approach to elicit information that other approaches cannot; and (d) its ability to lead to new approaches.

While monetary flows are typically used to follow interactions in human systems, "energy analysis is especially useful when imperfect markets or hidden subsidies distort prices and make economic analyses difficult" (Johnson et al. 1977, p. 374). Problems of relying on cash flow are exacerbated in populations loosely attached to national economies where tasks (particularly female and child labour) and exchanges are carried out without financial remuneration.

As Thomas et al. (1979) have noted, while no single factor can adequately explain the complexity of ecological and human systems, energy perhaps comes the closest. It is required by and underlies life processes, whether biological or social. At the very least, it serves as a common denominator in quantifying dynamic relationships within them and providing a basis for their comparison. Rather than an analytical end in itself, energy is a convenient starting point in the study of human-environment relationships. It is quantifiable, making testing of hypotheses possible - hypotheses which can be based on an extensive literature in energetics.

The study of energy flows in human groups is a one-variable approach, and thus lends itself to considerable criticism (Vayda & McCay 1975). The approach is valuable, however, since it serves as a broad organizing concept, within a limited range of reference. Its utility lies in the wide variety of phenomena - behavioural, ecological and biological - that influence or are influenced by the flow pattern. Caloric values can be assigned to most biobehavioural phenomena, and their influence quantitatively assessed.

The implications of energy-flow studies for human-population biology have been reviewed by Baker (1974). Figure 1 is a simplified diagram illustrating the energetic interconnections of sub-areas within physical anthropology. Questions of most interest to human biologists in relation to energy flow are: "1) how changes in energy availability affect biological traits and 2) how changes in biological traits affect the capacity of produce energy" (Baker 1974, p. 17). For example, if growth

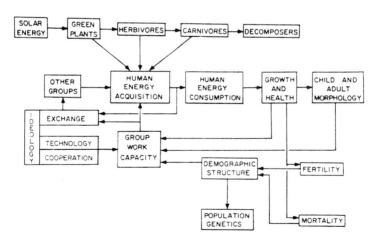

Figure 1. A simplified diagram illustrating energetic interconnections between areas of study in human biology (modified from Baker 1974).

rate of a population is decreased due to reduced protein availability, or if long-term debilitating disease (for example, tuber-
culosis or malaria) is widespread among adults, group work
capacity is adversely affected.  This requires an adjustment or
change somewhere in the system.

The neglect by physical anthropologists of the systemic
consequences of the phenomena they study has been noted by
Baker (1974): "Traditionally, human biologists have been concerned with some of these linkages.  However, almost all of the
studies have been on single components shown in figure 1. We
rarely look at the functioning of the whole system.  Rather, we
study the effect of the number of available calories on growth
rates in children, or the association between excess or deficient
calories and health measures, or how growth in children affects
adult size and health" (p. 20).  In contrast, Little & Morren
(1976) provide a good biocultural introduction of the utility of
human energetics.  A theoretical review can be found in Smith
(1979).  Methodology is covered in Jamison & Friedman (1974),
and Loucks & D'Alessio (1975).

In assessing the energetic consequences of alternative biobehavioural responses, several forms of energetic efficiency
can be calculated (table 1.)  Smith (1979) has reviewed the
relative merits of energetic efficiency formulae, emphasizing
differences between output-input efficiency and net acquisition
rate.  Bayliss-Smith, in his chapter in this volume, expands
upon these formulae.

## 4.  ANDEAN MODELS

The use of models to integrate both theory and data is
illustrated in figure 2.  Here, concepts derived from ecology,
demography and human biological sciences, capable of defining
general relationships between variables, are combined with
regional data on altiplano human-environmental systems.

The first effort to model such a system resulted in diagrammatic, energy-flow analysis of human adaptation to limited energy
availability.  This was carried out by Thomas (1973) in the
highland district of Nunoa (Department of Puno) in southern
Peru.  The study made use of extensive data collected by
Baker and colleagues from the Pennsylvania State University
(see Baker & Little 1976) who worked in the area from 1964
into the early 1970's.

Energetic relationships established in the diagrammatic,
human energy-flow model were built upon, and three simulation
models were created independently.  Each simulated the effects
of introducing some form of change anticipated to impact the

Table 1. Variables and parameters relating to energetic
efficiency

---

Variables

$E_a$ = energy acquired during some period of acquisition time

$E_e$ = energy expended in acquiring any $E_a$

$T_a$ = acquisition time for any $E_a$ (and the associated $E_e$)

Parameters

$E_n$ = net energy acquired = $(E_a - E_e)$

$R_g$ = gross acquisition rate = $(E_a/T_a)$

F = output/input efficiency = $(E_a/E_e)$

$R_n$ = net acquisition rate = $(E_a - E_e)/T_a$

---

area in the near future. A discussion of the contributions and
limitations of these models follows a brief description of the
Nunoan environment and human responses to it.

## 4.1.   The study area

The District of Nunoa lies at the northern extreme of the
Lake Titicaca drainage basin. Minimum elevation is 4000 metres,
and the zone utilized for food production extends up to the
frost desert at 5000 metres (figure 3). The population of the
District in recent times appears to be remaining stable at
approximately 8000 people. Quechua is the predominant language
and most residents make their living through agriculture and
pastoralism. Only a small number of mestizos were present in
Nunoa during the 1960s when it was intensively studied. These
individuals generally owned or managed haciendas which controlled much of the land and productivity of the area.
    Significant environmental conditions of this area have been
reviewed by Winterhalder & Thomas (1978). Those having a

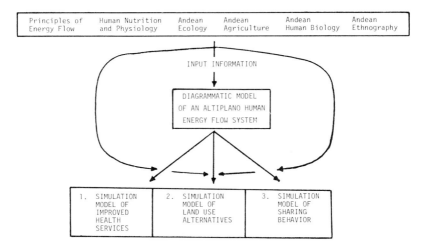

| Principles of Energy Flow | Human Nutrition and Physiology | Andean Ecology | Andean Agriculture | Andean Human Biology | Andean Ethnography |

INPUT INFORMATION

DIAGRAMMATIC MODEL
OF AN ALTIPLANO HUMAN
ENERGY FLOW SYSTEM

| 1. SIMULATION MODEL OF IMPROVED HEALTH SERVICES | 2. SIMULATION MODEL OF LAND USE ALTERNATIVES | 3. SIMULATION MODEL OF SHARING BEHAVIOR |

Figure 2. Information input into Andean simulation models.

primary influence on food production are: (a) environmental
heterogeneity in time, space and pattern resulting from rugged
topography; (b) poorly developed soils susceptible to erosion
and marginal availability of certain nutrients; (c) low temperatures with pronounced diurnal variation, and frequent and intense frosts which can occur in any season; (d) reduced oxygen and carbon dioxide tension, low absolute vapour pressure
and high background radiation; (e) a lengthy dry season,
irregular monthly distribution of precipitation and droughts
which may last for several years and which are unpredictable;
and (f) a biotic community with limited productivity spread
over wide regions.

While the lengthy dry season, soil conditions, and diurnal
temperature variations impose fairly constant limits on biota,
it is the irregular stressors such as frosts and droughts which
are primarily responsible for fluctuations in productivity from
year to year. The success of the human adaptive pattern is
largely explained by the wide range of biobehavioural responses
employed to decrease environmental perturbation and to increase
the quantity and variety of resources available.

These resources include a number of frost-resistant Andean
cereals and tubers in addition to three species of herd animal:

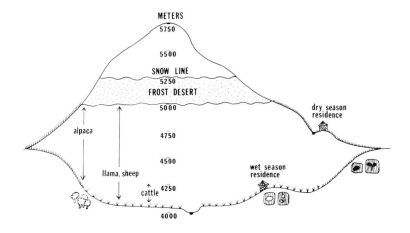

ALTITUDE LIMITS of CULTIGENS. NATURAL VEGETATION ANU HERD ANIMALS

Figure 3. Altitude limits of cultigens, natural vegetation and herd animals in Nunoa (from Winterhalder & Thomas 1978).

alpaca, llama and sheep. In terms of shared characteristics, they are capable of adjusting to and producing well in a variety of microzones, and are generally amenable to storage, transport and exchange with products of other regions. When possible, a multiple resource base consisting of items having different environmental tolerances and recovery rates is used. These often have nonconflicting schedules, and are spacially accessible to localized groups. Production techniques generally call for a dispersion of resources in time and space in order to avoid a simultaneous loss. This requires high mobility on the part of the productive unit, and dispersion of settlement pattern.

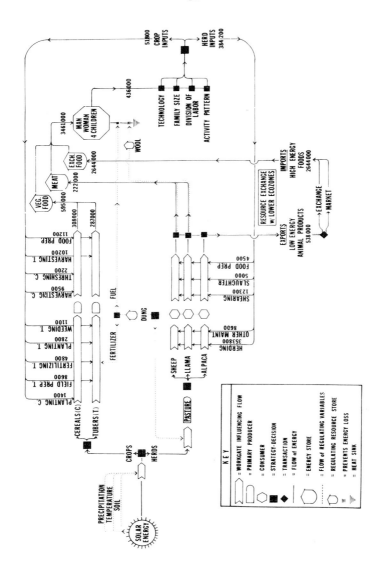

Figure 4.  Annual energy flow in kilocalories through a typical Nunoan family
(from Thomas 1976).

Finally, exchange between productive units of the same group serves to buffer the effects of localized resource loss and provides access to labour for tasks which the unit could not perform alone.     Exchange between groups residing in different zones has the same function in addition to providing essential resources not produced on the altiplano.

## 4.2.     Diagrammatic model: human energy flow

Limited capacity to channel food energy into the human population suggested that a number of energetically efficient biobehavioural responses might be found upon examining subsistence activities of rural families.     Goals of the investigation were therefore to determine levels of energy production, consumption and expenditure for families in order to describe their energy flow patterns, to assess the adequacy of their caloric intake, and finally to identify adaptive responses which facilitated adequate procurement and efficient utilization of food energy.

Analysis focused on productive and food-processing activities of families during an annual cycle, when harvests were considered normal.     Detailed attention was not given to exchange patterns outside the area or to the economic role of the hacienda.

The energy flow pattern for a modal nuclear family of four children is presented in figure 4.     Horizontal flows represent the amount of energy produced from local resources which ultimately reach the family.     Vertical arrows depict energy inputs into food production activities by family members.     Black squares indicate points in the flow system where decisions between alternative strategies have to be made.     Here one would expect that decisions which result in a higher energetic efficiency over a prolonged period would be more adaptive to limited energy availability.

Turning to the results of Thomas's study, it should be noted that many support observations long noted by ethnographers of altiplano groups.     Quantification of these data and their organization into a systemic model which focuses upon a single problem, however, provides a basis of comparison and experimentation with alternatives, not possible in more descriptive accounts.

Baker (1979) has summarized contributions of this work as follows.     Beginning with the human input, it will be noticed that the energy which was expended in crop production was influenced by a number of decision-making and structural characteristics of the family.     As indicated by kilocaloric expenditure, the vast majority of human energy expended in the

productive process went into animal production, rather than
agriculture. Agricultural inputs were, in fact, only about 1/7
as great as those in herding. The energy input for agriculture
was expended in the various activities of planting and harvesting. This activity yielded, at the end of a cycle, a total of
595 000 calories compared to an input of only 51 800. The
division of production by input gives an input-output energetic
efficiency of 11.3.

The energy-flow analysis in relation to herding is more complex. Herding behaviour requires the major input, while the
input into other animal-related activities is rather minimal. As
illustrated in figure 4, the output of herding was energetically
returned to the individual nuclear unit in two forms. First
was the meat itself, where caloric yield was quite low, providing many less calories than are actually expended. However,
538 000 kilocalories of herd production was traded or sold.
Thus, if the amount of calories directly utilized in the meat
and the amount of calories exported are combined, the yield is
about two calories for every calorie of input. However, the
export and sale of the animal products, including meat, wool
and skins, modified this relation substantially. From the exchange of the herding products for foods produced elsewhere,
the actual caloric yield rose to about six calories derived from
every input calorie. Even this value does not provide a complete measure of the energetic efficiency, since some of the
animal products were not exchanged directly for food, but
were sold for cash which was utilized in purchasing salt, cocoa,
tools, etc. If all of the money obtained by selling animal products is converted into purchased food calories, the energetic
efficiency is very similar to agriculture. That is, if all animal
products had been sold and exchanged for foods from sources
outside of Nunoa at the prices current during the study, herding would have yielded between 11–12 calories for every one
expended.

The generalizations which may be derived from this analysis
are many. First, this population obviously has a tight energy
budget, with a very limited excess. In addition, the system
probably could not have supported the population unless the
animal products from herding could be traded so that they
yielded a great caloric benefit. Without this benefit, the family
could obtain only two calories for every one put in, and would
probably not survive with this very low yield in relation to
energy input. By trading, the yield went from two calories
for each one expended to a yield of 11–12.

A second generalization is that the herding is a critical
regulator of the agricultural output of the community. This is
shown in the centre of figure 4, where it is indicated that dung
is critical for providing the fuel and fertilizer for this society.

Indeed, Winterhalder et al. (1974), in a more detailed analysis, showed dung to be the prime regulator of the amount of agriculture which can be carried out in the district. Unless other low-energy cost sources of fertilizer or fuel could be found, agricultural yield could not be increased without increasing the size of the herds. Since the sizes of the herds are regulated by the geographical area and quality of pasturage, this appears to have been a system whose yield could not be very substantially increased within its present technology.

In response to these restrictions on energy capture, it appears that the population had adjusted by a series of effective adaptations which allowed them to produce sufficient energy surplus to support the present population size. Among the adaptations which allowed the balance were the ways in which the productive labour was distributed by age and sex. Thus, the activities which could be accomplished by small individuals were assigned to children. If the same labour had been performed by adults, the energetic efficiency would have decreased significantly because the energy costs would have been higher. By the same token, certain productive activities are assigned to women. Ths use of women in the herding and agricultural activities also increased the energy efficiency compared to the use of male labour. Thus, any alteration in the age and sex distribution of the tasks within the system would reduce its efficiency, perhaps critically. In order to keep this labour system functioning, it was also necessary to export children as they became adults. Indeed, the population had a very high out-migration rate of about 2.5% per year in terms of permanent migration and an even higher rate of seasonal migration which reduced the caloric needs. This migration consisted primarily of young adults, and thus did not reduce the child labour force required to keep the system functioning.

Finally, it is argued that the heavy reliance on pastoralism was critical for this population to maintain itself, not only because it provided the necessary dung, but also because the herds represented as on-the-hoof storage of energy. The need for this reserve was important because of the year-to-year fluctuation in the potential productivity of agriculture. Long-term climatic records indicated that total crop failures must have been rather common occurrences in Nunoa. Thus, although additional land for agriculture was available, the results of heavy reliance on agriculture would have been disastrous in those years when the yield was nonexistent or very low. With the herd maintained, however, a buffer existed for food, allowing the population to survive in years when agricultural productivity provided a low energy yield.

In summary, the diagrammatic model of energy flow through

a modal Nunoa family, however simplistic, has raised numerous questions about the nature of this and other altiplano human-environmental systems. It has drawn attention to the indispensable nature of herd size in providing dung and products for exchange which are critical for the support of the population. In addition, it has emphasized the importance of different types of exchange and the inadequacy of the Nunoan data in this area. Quantification of levels and flows has in turn permitted experimentation with the efficacy of alternative solutions and comparisons between biological and behavioural responses.

The analysis indicated that, for whatever reasons, a number of important biological and behavioural responses fit expectations of what would be adaptive in an energetic sense. Obviously, not all behaviour conforms to this pattern. When high energy input results in no energy output, for example fiestas, it signals that other goals are important enough to counteract this pattern and deserve attention.

The principal problem with the diagrammatic model lies in a one-variable-at-a-time mode of analysis. While feedback is assumed, it does not become analytically tractable. Adaptiveness is therefore assessed at strategy decision points in the system. It is difficult, however, to comprehend how an energetically efficient alternative will affect a number of other responses in a precise manner.

As McRae (1979) has noted, interconnected adaptive responses have two characteristics. Firstly, they are engaged in a hierarchical fashion, meaning that priorities exist for activating a series of responses. In time of drought, for instance, people are likely to slaughter or sell some animals before moving out of the area. Secondly, responses influence each other positively and negatively. Hence, their cross-adaptive properties become important considerations in evaluating a response's effectiveness. These considerations limit the usefulness of analysing variables one at a time, and suggest that more integrated systemic approach, such as that provided by simulation modelling, could better approximate reality.

## 4.3.  Simulation models

Although simulation modelling of human-environmental systems is clearly no substitute for careful empirical research, the analysis combines concepts of systems dynamics, known relationships between system components, and a set of tools designed to translate system descriptions into formal models. These models, in turn, can explore the probable response of the entire modelled system to a variety of conditions, rather than

of only specific points in that system as the diagrammatic model
is capable of doing (Shantzis & Behrens 1972).

The Blankenship model: effects of improved health. The
first attempt at simulation modelling was carried out by James
Blankenship (Blankenship & Thomas 1977). It addressed a
type of change affecting many peasant groups throughout the
world as they become integrated into national economies and
modern health care. In the integration there are normally
three basic changes. Firstly, modern medicine is introduced,
drastically reducing mortality, particularly among children.
Secondly, either fertility control is introduced intentionally or
the population goes through a voluntary birth-restriction process. As a third or sometimes contemporaneous event, attempts
are made to increase the economic productivity of the population, often by improved agricultural methods.

Since rather complete data on the energy flow and the
demographic structure of Nunoa were available, it was possible

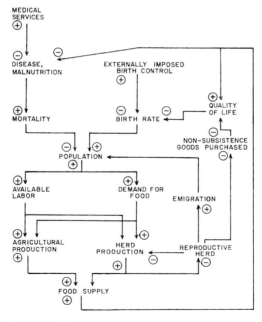

Figure 5. Causal and feedback relationships included in the Blankenship
model (from McRae, 1982).

to develop a mathematical model which simulated the demographic consequences of these changes within this small population.    In designing the model, productive capacity of the Nunoa region (represented in the diagrammatic model) was linked with demographic characteristics of the population.   This permitted an analysis of the probable effects of introducing various degrees of modern health care, and their systemic consequences, particularly the negative feedback loops influencing the region's carrying capacity and the population's quality of life (figure 5).

The model was intended to be simple, exploratory and general, applying to a wide range of peasant groups which employ mixed agricultural strategies and which are exposed to the aforementioned changes.   Consequently, a number of simplifying assumptions were introduced which do not reflect reality in antiplano groups.   Out-migration, which is an important factor among Andean populations, was not considered in the response repertoire.   Thus, the model pertains to groups having a fixed land base, whose members do not have or select the option of leaving.   Although the authors were well aware of the problems of assuming a carrying capacity, they have employed this concept as one such simplification.   These assumptions, however, do not seem to distort the generalizable utility of Blankenship's model in generating a set of testable hypotheses applicable to specific human-environmental systems. In sequence, four levels of prediction are discussed.

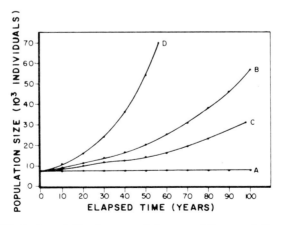

Figure 6.  Effect of mortality decrease of A, none of the population;
B, 0-5 year age groups; C, 6-85 year age groups; D, all age groups.
Unlimited resources assumed (from Blankenship & Thomas 1977).

The first examined only the question of what would happen if the mortality of the population was reduced with no other demographic or energetic modifications.  Figure 6 shows the predictions of this model.  If the mortality were reduced in all segments of society to the level found in industrial countries, Nunoa's consequent population growth, as shown by the line D, would be doubled in 20 years.  If, as is more common, the reduction in mortality affected primarily infants and young children, the doubling of the population would require about 40 years, as shown by line B.  Finally, as shown by line C, if, by some mechanism, mortality reduction were only to affect older children and adults, the doubling would take considerably longer.

The second-stage model considered how much and how fast fertility would have to be reduced in order for the population

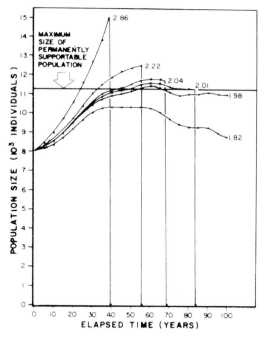

Figure 7. Effect of instant decreases in mortality and birth rates on population size.  Birth rate decreased from 6.67 children per female per lifetime to levels indicated on graph (from Blankenship & Thomas 1977).

to stabilize in numbers. For this model, it was assumed that
the mortality pattern resembled that of a modern population.
The results of this model are shown in figure 7. As can be
seen, an extremely drastic, immediate reduction in fertility
would have to occur for the population to stabilize in a reasonably short time. Indeed, the population would not stabilize
without a reduction from the present completed fertility of 6.7
children per woman to a fertility of two, and even this stabilization would require 60 years. The final size of the stabilized
population would be approximately 50% greater than the present
number. Stabilization at an earlier time would require a total
completed fertility considerably below two children per woman,
and even the most drastic reduction would be accompanied by
some increase in population size.

The next level of the model integrated the demographic projections with information from the energy-flow model. Based on
the model predictions, the total size of the population which
could be supported in Nunoa at the time of the study was calculated. This number was a little over 11 000 people. Most of
the additional support capacity came from increasing the emphasis on pastoralism and utilizing all of the cash return from the
trade system for the purchase of externally produced food. The
results are illustrated in figure 7. As can be seen, a population
with a birth rate above two would still exceed the support capacity of the system very rapidly. Indeed, even with fertility
as low as three children per woman, it would exceed the present
productive capabilities of the area in less than 20 years. This,
of course, barring any other circumstances, would lead to a
complete crash of the population because of starvation. Thus,
the survival of a low-mortality Nunoa population, without alteration of the technology and without changes in out-migration or
the prices of wool, requires a fertility below two. This reduction in fertility would still lead to a population with a poorer
standard of living than the one encountered at the time of the
study.

The final stage of the model examined the question of how
much the situation would be improved by increasing agricultural
yield. It would, of course, be theoretically possible to increase
this yield dramatically if improved hybrid crops were introduced,
and if fertilizer was made available from other sources. A new
model was therefore created on the assumption that total agricultural yield could be increased five-fold. The analysis suggested that a five-fold increase in the agricultural yield, without modifying herd production, would allow the technology to
support a total population of about 17 000. If the birth rate
did not drop in spite of the reduced mortality caused by the
introduction of modern medicine, even the five-fold increased

yield of agriculture would not stop the population from exceeding its energy production capacity within 20 years, and again disastrous consequences would occur. However, this simulation suggests that the birth rate, which finally came down to a bit over two children per woman, need not be reduced immediately. Instead, a stability could be reached in population numbers below the productive capacity of the district if the the rate was reduced from 6.7 to two within a 25-year time span. That is, the population would have to reduce its fertility fate from the present high one to the low one in a continuous manner over 25 years. Having done so, it would then reach the carrying capacity of the system within 60 years, but would not exceed it.

Let us reiterate that these models are not predictions of what is going to happen, but only predictions of what could happen if only the variables which have been included are modified. Thus, in the real world, the following is clear.

(a) Modern medicine introduced to Nunoa would be unlikely to produce the exact response predicted.

(b) The economic value of the herding products is not constant over time. If this were to increase in relation to carbohydrate food costs, the carrying capacity of the system in terms of numbers of people would certainly increase.

(c) The area is already experiencing a very high out-migration, and if the population continued to grow at a more rapid pace than at present, the out-migration would undoubtedly increase proportionately.

At one level, the model confirms a set of processes that seem intuitively obvious. That is, increasing health services should reduce infant and child mortality, and thereby increase population size. Furthermore, if fertility decline or out-migration does not compensate for lower death rates, and no economic growth were to take place, the quality of life in a material sense would also be expected to decline. Beyond this point, intuition does not help much, for it is very difficult to imagine how the changes in levels of the variables mentioned would affect others in a precise manner. Quantification allows one to perform this kind of experimentation with the system, and arrive at much more concrete observations about its operation.

These observations are useful in generating new research questions which modified forms of the model can address, or for which better empirical data can be gathered. For instance, the order in which government services such as public health, family planning, education (expected to increase out-migration), road building and agricultural improvement are introduced would very much influence population pressure on essential resources. Similarly, focusing on morbidity, one might identify those pathologies which are most disruptive to the productive capacity of the population and target these for priority attention.

The McRae model: land-use alternatives. This model, developed by McRae (1979), addresses another aspect of change prevalent in the Andean region where agrarian reform and attempts at agricultural specialization are expected to bring profound alterations in human systems. McRae compares the productivity and resiliency of three commonly utilized land-use alternatives, in order to assess their adaptability to the heterogeneous and unpredictable conditions of the altiplano. Land use alternatives considered are as follows:

(a) the 'standard system' consists of households engaged in farming and herding on their own or community land;
(b) the 'hacienda system' adds labour and herding obligations to the family's responsibilities of farming and herding its own animals;
(c) the 'alpaca-based system' specializes in alpaca herding without hacienda obligations.

Each version of the model is simulated under conditions representing both drought (stress) and non-drought (non-stress) situations. The non-stress simulation is examined over ten years under what might be considered optimal conditions. The environmental stress mimics the effects of a three-year drought with heavy crop losses and high herd mortality. In this situation, the simulation begins with a three year stress period in which the human population is provisioned with a season's worth of resources. A variant on this theme is considered, with the initial resource buffer amounting to a five-year accumulation of resources. For both of the simulations involving a stress period, the initial stress period is followed by a seven-year recovery period.

In contrast to the Blankenship model, McRae's emphasizes a considerably greater degree of realism. An impressive number of variables not derived from the energy-flow analysis have been entered into the model and are listed in table 2. These include not only the human demographic variables, but also the herd animal vital rates, factors controlling crop production, other income sources (handicrafts) and financial data. The major model components are given in figure 8. The simulation is largely based on two types of submodel: population projections and resource-allocation routines. The human and herd

**TABLE 2.** Principal Inputs and Outputs of McRae Model.

Input Variables

A. Simulation set up.
  1. Years the simulation is to run
  2. Herd age classes
  3. Human population age classes
  4. Human labour groups
  5. Work activities
  6. Resources demanded and supplied

B. Environmental stress routine.
  1. Survivorship reduction for herd animals
  2. Reduction in crop yield
  3. Beginning and termination points of stress period

C. Human population.
  1. Male and female population age structure
  2. Fertility schedule
  3. Sex ratio at birth
  4. Male and female survivorship schedule
  5. Level of background emigration
  6. Age/sex migration pattern

D. Herd population - three species, data for each.
  1. Male and female population age structures
  2. Fertility schedules
  3. Sex ratios at birth
  4. Male and female survivorship schedules
  5. Hacienda herd levels
  6. Initial proportion of herd sold and slaughtered

E. Animal product production.
  1. Wool yield, percent animals sheared
  2. Wool, meat, hide, and organ yields from slaughtered animals
  3. Live weights

F. Crop production
  1. Crop yields per hectare for three crops
  2. Crop reduction due to fertilizer limitations
  3. Seed application levels for three crops

## TABLE 2 continued

### G. Craft production.

1. Craft products produced per unit wool supplied
2. Proportion of craft products produced for trade

### H. Resource demand - six resources.

1. Annual money requirement per household
2. Annual wool demand per household
3. Annual fuel demand per household
4. Annual fertilizer demand per hectare
5. Energy and protein demand are calculated by the model

### I. Resource supply - twenty-five sources.

1. Initial levels of animal products, crops, crop seed, live animals, currency, animal dung, and craft products
2. Prices of resources supplied (sell) and demanded (buy)

### J. Land use.

1. Amount of arable land, total land
2. Household land demand - crops and pasture
3. Fallow period

### K. Resource allocation - labour.

1. Ranking schedule (priorities) for work activities
2. Preference schedule for labour groups
3. Work level requirements for each activity
4. Work availability for each age group

### L. Resource allocation - other.

1. Resource demand priority schedule
2. Preference schedule for resource supplies

### M. Energy flow, nutrition.

1. Energy and protein values of foodstuffs
2. Energy expenditure schedules by work group, work activity, and season of the year, and for non-work portions of the day
3. Protein requirements schedule by sex/age group

### N. Response routine

1. Extent to which pasture capacity can be exceeded before herd slaughter rate is increased.
2. Maximum allowable herd sizes
3. Minimum herd size before human out-migration begins.

models operate on a system of priorities to determine which resource demands are satisfied first, and a system of preferences
which determine the order in which different sources of supplies
are used to fill those demands. While the model is general in
the way it is constructed, it can be calibrated quite precisely
to represent a specific situation.

For purposes of model experimentation, a hypothetical community of 1000 persons is created having the same demographic
and behavioural characteristics as the Nunoa population of
herder-farmers described by Baker et al. (1968), Baker &
Little (1976), and Thomas (1973, 1976, 1978). McRae's models
were constructed with assumptions of no significant changes in
technology or human demographic behaviour, i.e., fertility,
mortality and migration.

In comparing the performance of the three formulations of
land use, it is instructive to look at human and herd population
dynamics over the ten-year run of the simulation. Under an

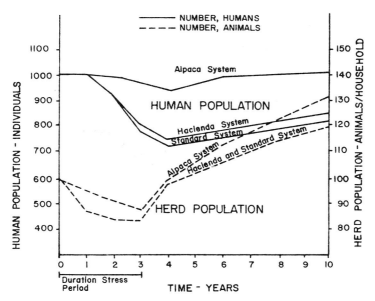

Figure 9. Human and herd population trends for three variants of the production where the population, under an environmental stress scenario,
lacks a substantial initial resource buffer. The environmental stress
is produced by a three-year drought.(from McRae 1982).

environmental stress scenario, the time-course of the human population numbers is a partial indicator of the degree to which the population is able to withstand a period of environmental stress.  During stress periods, reduction in herd size was brought about either because herd animals had to be sold to acquire needed resources or because of increased herd mortality.  While human population levels reflect the impact of a period of stress on one of the most important components of the production system, herds, they do not indicate other impacts, such as disrupted social relations.  Despite this, they remain useful comparators for alternative systems.

Figure 9 shows human and herd trends to a prolonged three-year drought (stress period) and a subsequent recovery period.  It can be observed that the herd population in all three systems drops during the stress period.  Herd mortality is elevated 5% for alpaca and 10% for sheep, whereas crop yield is reduced by 80%.  The effect on the human population is

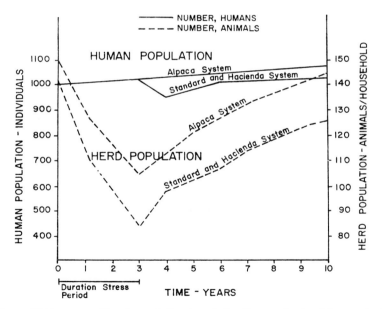

Figure 10.Human and herd population trends for three variants of the production system under an environmental stress scenario with a substantial initial resource buffer.  The environmental stress is produced by a three-year drought (from McRae 1982).

marked, decreasing from 1000 to a low of 934 in the alpaca system, versus 719 and 742 in the standard and hacienda systems. The alpaca system is not only losing fewer animals, but is also producing more alpaca wool which, based on mid-1960s prices, was almost six times more valuable than sheep wool. As a traded commodity, it was therefore commensurately more effective in procuring resources to make up for those lost in the drought.

The above scenarios are of a population undergoing a severe drought with only about one season's worth of resources in storage. What might happen if they had a greater resource buffer? The results of a simulation experiment are given in figure 10, where five non-stressed years precede the onset of the three-year stress period in order to provide such a buffer. The primary buffer is obtained through the increasing herd sizes, which become large enough to enable them to be reduced dramatically during the stress period. Relatively few people are forced out of the system in the case of the standard and hacienda systems, and none in the case of the alpaca system.

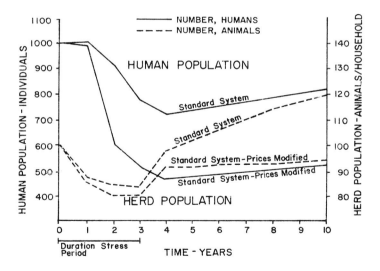

Figure 11. Human and herd population trends using two versions of the standard production system. In one of the versions, key prices are changed, having the effect of increasing the amount of alpaca wool that must be exchanged to acquire a given amount of energy in wheat flour (from McRae 1982).

Since trading animal products ia a major component of the production system, the role of prices is a significant factor in the relative support capacity of the simulated production systems. An experiment to examine how significant price changes might be was performed by lowering the price of alpaca wool and raising the price of wheat flour. Both events were programmed to occur during a stress period. The results of these changes upon the system's buffering capabilities are shown in figure 11. The system being tested was the standard one. Price modified curves resulted from decreasing the price of alpaca wool by 38% and raising the price of wheat flour 50% (in the model, wheat flour represents the primary source of energy coming from the lower areas). Sheep were eliminated from the system by being sold and the human population was reduced to 463 in year four of the simulation. This scenario demonstrates the heightened effect of combining an economic with an environmental stressor, and suggests that the human support system provided by herding is very vulnerable to this type of stress pattern. It also underlines the importance of storage in a variety of forms as an adaptive strategy.

With the exception of impact on pasture, the alpaca system seems to be superior to the others for most measures examined. However, this conclusion may be modified by the facts that the alpaca is productive only in the upper puna zone and that the effect of interspecies variability in terms of impact on pasture was not considered. Only when the extent of accessible dry-season pasture in the upper puna in the Nunoan region and the impact of different species on the pasture are known will it be possible to reach more definitive conclusions by entering these variables in the model. Likewise, no consideration was given to the incidence and severity of herd pathologies which might increase the risk of relying upon a single species.

The results presented above are seen as preliminary, serving primarily as a focus to further research and discussion. Further testing with the present model as well as further model development are needed and anticipated. Data upon which the model is based were collected primarily in the middle and late 1960s. Since that time, agrarian reform has eliminated the hacienda system, replacing it with the two other alternatives. This provides an excellent opportunity to validate predictions of the model. Slaughter and sale of hacienda herds, for instance, significantly reduced the Nunoa herd population, suggesting that dung for fuel and fertilizer may be scarce in the region. Conversely, removal of hacienda obligations has freed labour. This could be used for building up a family's storage capacity well beyond that possible under the hacienda system. As we have seen, this might considerably reduce the effects of drought and unpredictable market prices.

The Oak Ridge model: sharing behaviour alternatives.  The unpredictable nature of environmental problems on the altiplano presents conditions where a single family cannot rely solely upon its own efforts or stored resources, and thus must depend upon other individuals within and beyond the group.

While such dependency can take a variety of forms, reciprocal relationships based on kin and fictive kin are emphasized in rural areas (Alberti & Mayer 1974).  One advantage of forming co-operative units with relatively strong and long-lasting bonds, such as kin co-operation, is the extent to which generalized reciprocity can operate.  Repayment in kind or within a particular time period is not emphasized as it would be under a more formal arrangement.  Such an arrangement provides considerable flexibility for a household encountering a sequence of bad luck, since it demands only that the household contribute to the co-operative unit what it can afford.

In evaluating generalized reciprocity as an adaptive strategy, Weinstein et al. (1979) from Oak Ridge National Laboratories have developed a detailed and versatile stochastic model, which considers each individual in a hypothetical altiplano group separately.  This differs from the Blankenship and McRae models, which are deterministic in nature, establishing key relationships in a fixed rather than a probabilistic manner.

The goal of the Oak Ridge effort is to track individuals within the population in order to investigate adaptive phenomena at the individual, family and kin-group level of organization.  This is not possible with the other two models, which are designed to consider adaptive phenomena at the population level.

The Oak Ridge model is composed of three submodels: the family, agriculture, and herds (figure 12).  The family submodel "calculates caloric requirements for each family based on the sex and age of its members, and keeps track of their births, deaths and marriages as they change over time.  The agricultural submodel calculates energy expended and produced per year for seven crops.  The effects of temperature, precipitation, potential evapotranspiration and occurrences of frost and hail which affect productivity are stochastically introduced.  The herd submodel maintains an inventory of three herd species owned by the family and simulates herd demographic behavior including slaughter and sale of animal products.  Sharing of surplus food among extended family members is also handled by this submodel" (Weinstein et al. 1979, p. 517).

In examining resource sharing, the simulation commences with a population of 100 nuclear families equally divided into 20 extended families (designated as clans).  The model then is run for 250 years, or approximately 12-15 generations, with clan-size inventories taken every 50 years.

Preliminary results suggest that extended families which expand rapidly do so by accumulating a large resource base available for sharing. This, in turn, leaves fewer resources available for other extended families blocking their expansion. In spite of a significantly restructured distribution of families in the course of the simulated period, the number of families and the total population size remain fairly constant, indicating a stable population (Weinstein et al. 1979).

In order to examine the role of sharing more directly, three cases of resource sharing are compared: no sharing between families, sharing within extended families or clans, and sharing between all families. Since models with stochastic components must be repeatedly run to determine the mean and variance of the resulting variables, two 250-year simulations were made for each case, starting with the same initial population and weather

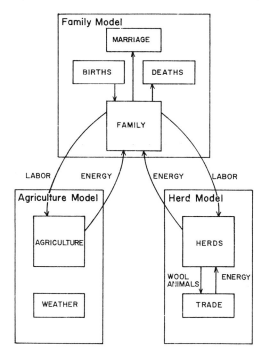

Figure 12. Major models, sub-models and linkages in the Oak Ridge model (from McRae 1982).

conditions. The random-number sequence for the stochastic models was changed for each repetition.

The population is subjected to stress through an energy deficit resulting from crop loss, based on the probabilistic occurrence of hail or frost. With restrictions on the amount of energy that can be imported through trading animals and animal products due to, among other reasons, a finite herd size, energy deficits also result on a probabilistic basis.

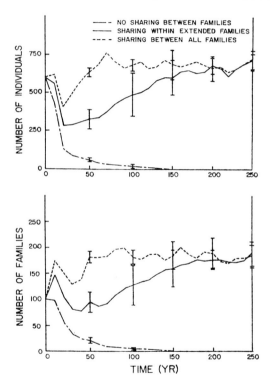

Figure 13. Effect of resource sharing on population size and the number of families through time. In the model, an initial population of 100 families (644 individuals) is distributed among 20 clans. Because this is a stochastic model, multiple runs are performed, and means and variance (vertical bars represent one standard deviation) plotted. This figure was kindly contributed by D. Weinstein (from McRae 1982).

There is a negative feedback loop linking population with energy deficits. Thus, births are reduced and deaths increased due to energy deficits. The effects on families and individuals are shown in figure 13. The results suggest that families cannot survive alone, isolated from other families. This result is corroborated by results from the McRae model, which assumes sharing of all resources (including labour) within the community of 1000 persons.

While the long time-span over which the Oak Ridge simulation runs brings into doubt the realism of assumptions that future production relationships will remain the same, the model nevertheless provides an opportunity for anthropological experimentation not possible in the real world. For instance, it would be of interest to compare sharing potential between a group having a bilateral kinship system, where relatives are claimed on both sides of the family, as opposed to one with a matrilineal or patrilineal system. Likewise, one could trace the spread through the population of advantageous genes, such as those affecting increased working capacity, using different mating systems. The culmination of deleterious genes, affecting susceptibility to respiratory disorders and reduction in working capacity could also be followed.


5. SUMMARY

In assessing the role that modelling can play in anticipating change in the Andes, three simulation models have been considered. All of these share a data base taken from the diagrammatic, human energy-flow model, yet each has added information relevant to a specific type of change. As a result, they differ substantially in the number of relationships, the nature of these relationships between variables, and time periods considered.

It should be kept in mind that these represent three independent attempts to assess the utility of this approach, and hence should be regarded together as a first generation of altiplano models. The Blankenship model, for instance, emphasizes generality, or a set of feedback relationships between health, population growth and limited resources assumed to apply to human groups residing on and beyond the altiplano. Many simplifying assumptions, such as the absence of out-migration, however, make the model unrealistic in terms of what we know to exist among highland populations.

McRae's model on land-use alternatives approaches the reality of the altiplano much more closely. It incorporates considerable information from the region. Likewise, it assumes that observed relationships between variables are transitory, and that the analysis would lose validity after 10-15 years. In spite of this attention to realism, however, its design is general enough to allow it to be applied to a wide variety of groups reliant upon crops and animals under different conditions.

The Oak Ridge model, which was designed to consider stochastic or probabilistic relationships between some of its variables, introduces an added dimension of reality mimicking the unpredictable environment. While its long time-stage makes the analysis somewhat unrealistic, in times of rapid change, it becomes an excellent tool for exploring consequences of long-term genetic or social change which could not be tracked by other means.

In short, the three models address various degrees of generality, reality and precision. Of necessity, all have simplified reality of altiplano human-environmental systems to some extent, and have introduced biases derived from the Nunoa data base. What is apparent, however, is the analytical diversity of these first-generation simulation approaches, and their potential of being incorporated into a much more sophisticated modelling effort.

Together they have provided insights into the operation of altiplano systems, identified areas sensitive to change where more precise data are needed, and analysed effects of alternative strategies upon system operation. The extent to which alternatives judged as better by simulation modelling are generally acceptable by other scientists, planners and people residing in the system is open to debate. This, in fact, is one form of validation of the model.

In conclusion, modelling techniques seem particularly appropriate in the Andes, where rapid change is coupled with a diversity of human-environmental systems and a limited data base. Clearly, the time-frame within which most changes will impact these systems is not sufficient to collect detailed data on each set of conditions. We do, however, have information on a general set of relationships operating in the Andes, and detailed data from selected areas. Combining these within a modelling approach whereby it is possible to change conditions and responses to fit specific situations may make most efficient use of these data. The model under these circumstances cannot be used so much as a predictive device, but as one of self-instruction in some potential consequences of alternative responses and policies to change.


6.  POSTSCRIPT

Since a general optimism prevails throughout this chapter

concerning the utility of human energy-flow analysis, one is
referred to Philip Burnham's chapter for a rather contrasting
point of view. Here, he attempts a 'nihilistic' critique of
Thomas's initial Nunoa energy-flow study, suggesting irrepar-
able problems with data collection and analysis. Not only is an
adaptive interpretation based on energetics found to be overly
restrictive, but so is the human-adaptability framework and the
positivist-functionalist, tradition it reflects. Simply put, he is
asking for greater precision in the data base, a higher degree
of reality in the number of parameters included in the model,
and a broader or different theoretical approach. Throughout
this seeming unending barrage of negativism a constant com-
plaint is the study's lack of attention to social relations within
and beyond Nunoa.

Given Burnham's uncompromising position, it is appropriate
to start with some points of agreement. In attempting to link
together components of a human-environmental system which
relate to energy production, consumption and expenditure, precise measurement of all variables is not always possible (Jamison
& Friedman 1974). This is inevitable when few investigators
are available to conduct such a broad-based study. Nevertheless, two individuals working for 14 months in Nunoa did find
it possible to obtain a general quantitative description of energy
flow through the system. Obviously, a large research team
would have led to more effective data collection, and previous
work in the region on human growth, work physiology nutrition
and demography was extremely helpful in evaluating findings
and extending conclusions.

A second point of agreement is that much has changed in
the way in which we view human ecology and energy flow since
the initial Nunoa study was conducted in 1968 and made available in monograph form in Thomas (1973). As the human-
ecology approach has been used and abused, its practitioners
have developed increasing methodological and theoretical sophis-
tication, in addition to a better sense of its limitations. We
have tried to go beyond evaluating systems in terms of a single
flow, by considering multiple flows and constraints. Likewise,
simple indicators of adaptiveness, such as an energetic ratio,
have been replaced by a number of more sensitive measures of
energetic efficiency. Universalistic laws concerning the energetic organization of human affairs seem never to have been
seriously considered by even the earliest human energy-flow
researchers in anthropology or geography, who were more
interested in interpreting local systems. In fact, energy-flow
analysis as a method is quite separable from an adaptive framework or any other theoretical stance.

Early human energy-flow studies produced in the 1960's

and early 1970's have been rightly criticized for their over-emphasis on closed-equilibrium systems instead of open-dynamic ones, and for description of modal flow patterns rather than addressing variability.   However, narrow and naive though these studies may seem from our present vantage point, certain simplifications were probably necessary to test the utility of the approach.

We are therefore in agreement that the initial study re-flected the state of the art at that time and could have been more precise and broader in scope.   The authors would, how-ever, maintain that the study was quite adequate for taking the first exploratory steps.   It does not require a great degree of faith to surmise that Thomas was aware of external political eco-nomic relationships influencing local Nunoan conditions, or of class and ethnic interactions affecting self-sufficient responses at the household level, or that the human system was under-going change.   These things are all too apparent in the Andes.   Hopefully, our present chapter gives the impression that we are concerned with these factors, and has suggested that energy-flow analysis can go beyond the initial simplifying assumptions to examine open-dynamic systems.

What disappoints us most in the critique is that Burnham has expectations of the research process which are unrealistic, and expectations of elucidating a set of relationships which dif-fer considerably from the initial study's design and goals.   Since these expectations aren't met, we are advised that this area of study has little to offer.   In a discipline such as anthropology, where understanding human biological and cul-tural phenomena from a systemic point of view is of central con-cern, it is discouraging to find that paths of inquiry are so easily dismissed.

Concerning unrealistic expectations of the research process, it has been pointed out in our discussion of models that simul-taneous emphasis on precision, reality and generality is usually not possible.   Since models are designed to examine certain relationships within systems, they cannot be all-inclusive if we desire to use them for testing purposes.   In proposing a new model, such as energy-flow analysis, it seems important to sim-plify the problem as much as reality permits in order to assess its explanatory value.   In this simplification process, assump-tions are made that are unrealistic, but which make the analysis feasible and which are thought not to seriously distort the find-ings.   We are fairly explicit as to what these are in the present chapter.

It is important to stress again that the Nunoan energy-flow study, carried out over a decade ago, was an exploratory effort, as were Rappaport's and other initial studies done

around this time. These were independently influenced by
energy-flow analysis in ecology and were applied to groups
with the hope of better defining human systems. If the use
of an adaptive indicator such as an energetic input-output ratio
to evaluate the relative benefit of responses appears overly restrictive, this was intended in order to test the value of the
measure. The fact that many biological and behavioural responses having a high energetic efficiency were frequently utilized suggested that energetic relationships had some bearing
on system organization and that the model was worthwhile pur-
suing. Although not reported here, the energetic data have
enabled us to take a closer look at the interplay between biological and behavioural responses to multiple stressors existing
at high altitude.

Unfortunately, Burnham seems only to have read a summary
article (Thomas 1976) of this initial work, which omits much of
the rationale of investigating the flow of a single variable
through a human biobehavioural system. The intent was to
explore relationships between biological and behavioural responses to local environmental conditions, and not political economic
issues. Research focused on subsistence needs at the individual and household level, and was designed to build upon previous work in human adaptability done in Nunoa.

In summary, the utility of energy-flow analysis lies in its
ability to quantify relationships between principal components
of human systems. Evaluation of such an approach is determined by (a) the importance of energy in explaining systemic
relationships; (b) the ability to measure the variable; (c) the
ability of the approach to elicit information that other approaches cannot; and (d) its ability to lead to new approaches.
The reader, of course, must make this judgement. Like other
currencies which are used to track the operations of a system,
energy-flow analysis serves primarily as a research tool.
Since it is not linked to a particular paradigm, we should be
able to track the dynamics and consequences of Nunoa's interaction with a national or world system, or of class/ethnic competition for resources within the region, as well as household
responses to drought. Rather than being an analytical end in
itself, energy-flow analysis serves as one convenient starting
point in understanding the complexity of human systems in
which social relations play a primary role.

Hopefully, this response has convinced Burnham, and
others who share his opinions, that energy-flow analysis has
some utility and could even be employed to address the very
relevant questions he is concerned with. Taking jabs on paper,
as we have done, at one another's work and opinions is relatively easy and is not always very productive. In this case,

it drives modes of analysis and interpretation apart which could potentially contribute much to each other. We suspect that had this debate been more than a paper confrontation, our ideas on where to take future energy-flow studies might be rather close to Philip Burnham's.

## 7. ACKNOWLEDGEMENTS

Funds to attend the SSHB Special General Meeting were generously provided by the Society to R. Brooke Thomas. Special thanks are due to Geoffrey Harrison and Tony Boyce of of the Department of Biological Anthropology at Oxford for their efforts and hospitality.
This paper has drawn from reviews of Andean modelling efforts by Baker (1979) and McRae (1982), and energy flow analysis in Thomas et al. (1979).

## REFERENCES

Alberti, G. & Mayer, E., 1974, Reciprocidad e Intercambio en los Andes Peruvanos (Lima: Instituto de Estudios Peruanos).
Baker, P.T., 1974, The implications of energy flow studies on human populations for human biology. In Energy Flow in Human Communities, edited by P.L. Jamison and S.M. Friedman (University Park Pa.: Human Adaptability Co-ordinating Office US/IBP and SSRC), pp. 15-20.
Baker, P.T., 1979, The use of human ecological models in biological anthropology: examples from the Andes. Coll. Anthropology, 3, (2), 157-171.
Baker, P.T. & Little, M.A. (editors), 1976, Man in the Andes: A Multidisciplinary Study of High Altitude Quechua (Stroudsburg, Pa.: Dowden, Hutchinson Ross).
Baker, P.T., Escobar, G., DeJong, G., Hoff, C.J., Mazess, R.B., Hanna, J.M., Little, M.A. & Picon-Reategui, E., 1968. High Altitude Adaptation in a Peruvian Community (Pennsylvania State University, University Park, Pa.: Occasional Papers in Anthropology No. 1).
Blankenship, J.C. & Thomas, R.B., 1977, Demographic impact of introducing modern medicine to a subsistence-level agrarian population: A Simulation. Environmental Management, 1 (5), 401-417.
DeGreene, K.B., 1973, Sociotechnical Systems. Factors in Analysis, Design and Management (Englewood Cliffs, N.J.,: Prentice Hall).

Forrester, J.W., 1968, Principles of Systems, 2nd Ed. (Cambridge: Wright Allen Press).

Holling, C.S., 1973, Resilience and stability of ecological systems. Annual Review of Ecological Systems, 4, 1-23.

Holling, C.S. & Chambers, A.D., 1973, Resource Science: The Nurture of an Infant. Bioscience, 23, 13-20.

Jamison, P.L. & Friedman, S.M. (editors), 1974, Energy Flow in Human Communities (University Park, Pa.: Human Adaptability Coordinating Office, US/IBP and SSRC).

Johnson, W.A., Stolzfus, V. & Craumer, P., 1977, Energy conservation in Amish agriculture. Science, 198, 373-378.

Lee, D.B., 1973, Requiem for large scale models. A.I.P. Journal, May, 1973, pp. 163-178.

Levins, R., 1966, The strategy of model building in population ecology. American Scientist, 54 (4), 421-431.

Little, M.A. & Morren, G.E.B., Jr, 1976, Ecology, Energetics, and Human Variability (Dubuque, Iowa: W.C. Brown).

Loucks, O.L. & D'Alessio, A., 1975, Energy Flow and Human Adaptation: A Summary. Office of Ecosystem Studies, The Institute of Ecology. Summary of a workshop on Energy Flow Through Human Communities. L. Harris and B. Thomas, organizers. University of Florida.

McRae, S.D., 1979, Resource Allocation Alternatives in an Andean Herding System: A Simulation Approach (Cornell University, Ithaca, NY.: Ph.D. Thesis). Published in Dissertation Series of Latin American Studies Program, No 82, Cornell University, Ithaca, NY.

McRae, S.D., 1982, Human ecological modelling for the Central Andes. Mountain Research and Development, 2(1), 97-110.

Shannon, R.E., 1975, Systems Simulation, the Art and Science (Englewood Cliffs, N.J.: Prentice Hall).

Shantzis, S.B. & Behrens, W.W., 1972, Population control mechanisms in a primitive agricultural society. In Toward Global Equilibrium: Collected Papers, edited by D.L. Meadows and D.H. Meadows (Cambridge, Ma.: Wright Allen Press).

Smith, E.A., 1979, Human adaptation and energetic efficiency. Human Ecology, 7, 53-74.

Thomas, R.B., 1973, Human Adaptation to a High Andean Energy Flow System (Pennsylvania State University, University Park, Pa.: Occasional Papers in Anthropology, No. 7).

Thomas, R.B., 1976, Energy Flow at High Altitude. In Man in the Andes, edited by P. Baker and M.A. Little (Stroudsburg, Pa.: Dowden, Hutchinson & Ross).

Thomas, R.B., 1978, Effects of change on high mountain human adaptive patterns. In High Altitude Geoecology, edited by P.J. Webber (Boulder, Colorado: Westview).

Thomas, R.B., Winterhalder, B. & McRae, S.D., 1979, An anthropological approach to human ecology and adaptive dynamics. Yearbook of Physical Anthropology, 22, 1-46.

Vayda, A.P. & McCay, B.J., New directions in ecology and ecological anthropology. Annual Review of Anthropology, 4, 293-306.

Walker, B.H., 1973, An appraisal of the systems approach to research on and management of Africal wildlife ecosystems (Symposium on Wildlife Conservation and Utilization in Africa, Pretoria, June 1973. Mimeo. 7 pp.).

Watt, K.W., 1968, Ecology and Resource Management (New York: McGraw-Hill).

Weinstein, D.A., Brandt, C.C. & Shugart, H.H., 1979, A computer simulation of energy flow through individuals, families, and extended families of the high altitude Quechua. In Changing Energy Use Futures, edited by R.A. Fassolare, and C.B. Smith (New York: Pergamon).

Winterhalder, B., Larsen, R. & Thomas, R.B., 1974, Dung as an essential resource in a highland Peruvian community. Human ecology, 2 (2), 89-104.

Winterhalder, B. & Thomas, R.B. 1978, Geoecology of Southern Highland Peru: A Human Adaptation Perspective (University Of Colorado, Boulder, Institute of Arctic and Alpine Research: Occasional Paper No. 27).

# ENERGY USE, FOOD PRODUCTION AND WELFARE:
## PERSPECTIVES ON THE 'EFFICIENCY' OF AGRICULTURAL
## SYSTEMS

T. Bayliss-Smith

Department of Geography,
University of Cambridge,
Downing Place,
Cambridge CB2 3EN

## 1.    INTRODUCTION

Who are the world's most efficient farmers?    This is a question
which can be tackled in an objective manner using energy analysis, but the answer that emerges will depend entirely on how
broadly or how narrowly we define a farmer's system of agricultural production, and secondly on how selectively we view
the various categories of energy flow.    If we adopt an ecological
perspective, a number of indices are available which portray
gross or net efficiencies in the conversion by plants of solar
energy into chemical energy.    But if we adopt the standpoint
of the economist, efficiency implies something quite different:
ecological processes are taken for granted, and what matters is
the production of food energy per unit of land, or of labour, or
of other inputs.    The sociologist, on the other hand, will point
out that in human terms some farming systems generate more
satisfaction and less inequality than do others.    Policies which
generate more farm output through capitalist amalgamation or
communist collectivization may not create an agrarian society
which is efficient in maintaining the quality of life for its members.    Such inefficiencies do, of course, spill over into other
sectors of society, as dissatisfied farm workers or landless
peasants migrate to urban areas in search of improvement.
This paper cannot possibly cover all these aspects of efficiency.
I do, however, suggest that an energy approach enables us to
quantify most material aspects of efficiency, by providing data
from which we can calculate various measures of performance.
Five particular indices are proposed, which taken together do
provide an accurate summary of the ecological, economic and
sociological dimensions of the problem.    I illustrate the use of
these indices by means of seven case studies of farming systems
which cover almost the entire gamut of variation in agrarian
adaptations, past and present.    In the first part of the paper,
five efficiency indices are defined and discussed.    Subsequently,

data for the seven farming systems are presented, and some implications are suggested.


## 2.    DEFINING AGRICULTURAL SYSTEMS

If we are to consider an agricultural system as a social entity rather than as an ecosystem, and if we wish to study the energy flows that enable that entity to function, at least three definitions of the system are possible (figure 1).

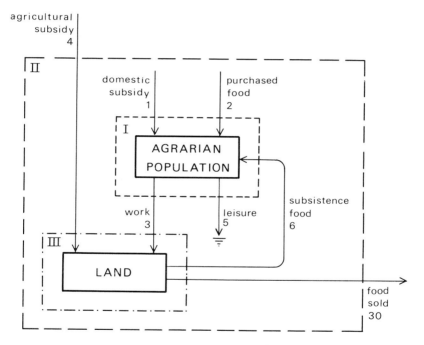

FIGURE 1.    Three definitions of 'agricultural system', showing different boundaries delimiting energy inputs and outputs.

The narrowest view would be to consider the energetics of the agrarian population (System 1 in figure 1).    Research would focus on the expenditure of energy by individuals,

whether in work activities or in leisure, as made possible by
energy intake derived from food produced, exchanged or purchased.   An extension of the study might include a consideration
of the domestic subsidies of energy (firewood, electricity, machines) needed to enable the population to maintain its lifestyle.

Such a view does not discriminate at all between different ways
of farming, except to the extent that some agricultural systems
require more manual labour than do others. The various energy
flows in farming itself have been analysed at two scales.   At the
larger scale, System II (figure 1) represents the agricultural
system as viewed by commercial interests and, all too often, is
that which is portrayed in official statistics.   The only output
regarded as significant is the component of production which is
sold in the market economy, and the only inputs known about
are the major categories of energy subsidy:  fertilizers,
irrigation water, machinery, fuel and power supplies.   Both the
work component and the subsistence sector are ignored.

The third definition of agricultural systems, System III, also
focuses on the farm, but considers work inputs and subsistence
outputs as well as the products that interest the outside world.
It would be more logical to include also as inputs to the farming
system imported foods and domestic energy subsidies used by
the agrarian community, but in the present study these are
excluded because of lack of data, and in order to conform with
existing conventions of energy accounting.   In Leach's words,
"no charge is made for the energy required to support the personal or domestic life of the farmer or farm workers since these
are of no relevance to the business of producing food" (p.60).

Using the nominal values shown in figure 1 for these various
categories of input and output, we have three systems which
vary considerably in the character and magnitude of the principal
flows.

System I (agrarian population)
    Inputs = food consumed + domestic subsidies = 9
    Outputs = work + leisure = 8
    Output/input ratio = 0.89

System II (commercial sector)
    Inputs = agricultural subsidies = 4
    Outputs = food sold = 30
    Output/input ratio = 7.50

System III (farm)
    Inputs = work + agricultural subsidies = 7
    Outputs = subsistence food + food sold = 36
    Output/input ratio = 5.14

We are concerned in this paper only with type III systems, together with the welfare of the farming populations that control
them.


## 3.   INPUTS INTO AGRICULTURAL SYSTEMS

We can divide the inputs into our system into the energy of
human labour and externally derived inputs which subsidize that
labour, notably machines and their power sources, fertilizers and
pesticides.    These 'energy subsidies' all in turn require energy
for their manufacture, marketing and distribution, and therefore
involve society in an energy cost.    The excessive dependence
of industrial societies upon these energy subsidies for its food
production has been highlighted by oil shortages in the last decade, and by a growing belief in an imminent 'energy crisis'.
The simplest way of expressing the work input into agriculture
is in terms of hours of effort per week per worker.    On the
other hand, if we want a direct comparison with other inputs,
and with outputs, we must convert the hours of work into an
approximate energy equivalent.    The different subsidies to
human effort are so varied anyway that unless we calculate an
energy equivalent, any comparison at all between different agricultural systems is difficult.    In this paper, both hours of work
per worker and the amount of energy thereby expended are used
in different contexts, according to which is the most revealing
way of expressing efficiency.


## 4.   INDICES OF OUTPUT - FOOD ENERGY YIELD

The first measure of output efficiency that we shall consider
is the familiar yield measure:  annual output of net food energy
per unit area.    Crop yields in any particular year will be controlled by a multitude of factors, natural and cultural, but if we
calculate instead the mean yield over an indefinite number of
years, the frequency of cultivation becomes one of the most important controls.    Areas of low population density occupied by
pre-industrial societies typically practise shifting cultivation if
it is ecologically feasible.    Forest fallows in the tropics are often
of decades in duration, so that in the long term food energy
yields per hectare are greatly reduced.    On the other hand,
agricultural systems also exist in the tropics, where perennial
cultivation is practised, and here substantially higher energy
yields than are typical of temperate farming can be found.

Data summarized by Leach (1976), and by Pimentel & Pimentel (1979), indicate the following range of variation for farming systems, classified according to the extent of energy subsidies (mechanization, fertilizers, etc.) involved in production.

| | |
|---|---|
| Pre-industrial crops: | 1—50 (GJ/ha)/y |
| Semi-industrial crops: | 15—60 (GJ/ha)/y |
| Full-industrial farms: | 1—85 (GJ/ha)/y |

'Pre-industrial' signifies that fossil-fuel inputs are zero or very small, below 10% of the total; 'semi-industrial' are systems where fossil-fuel-derived inputs comprise 10—95% of the total; in 'full-industrial' systems, the human energy input is below 5%, and generally so small as to be disregarded.

Whereas in pre-industrial systems the range of yields shown above is to be explained mainly by variations in the length of fallow, in the other systems the application of industrial technology (i.e. fertilizer, etc.) enables land to be kept in perennial cultivation.   Here, the principal cause of variation in yield is the extent to which farming systems produce animal products rather than crops.   Sheep farms in the UK, for example, have food energy yields similar to those of shifting cultivators in tropical forests.   UK farms producing eggs and poultry are similar in energy output to African farms in savanna regions growing maize, groundnut and millet on land that is only used one year in two.   In both cases, the difference between the systems lies in levels of input required rather than output achieved.


## 5.   GROSS ENERGY PRODUCTIVITY

Another traditional economic measure of efficiency is output per worker or per unit of work.   In this paper, this index will be defined as the amount of edible energy (animal feed included) that is produced per person per day, averaged out over the entire agricultural population directly supported by the agricultural system.   For pre-industrial societies, this definition enables us to consider the entire agrarian community, including old persons and children who make occasional but sometimes significant contributions to agricultural work.   Ideally, one should also include in the 'agrarian population' persons indirectly supported.   However, where agrarian societies have developed strong links with neighbouring populations, for marketing produce or for purchasing goods and services, it becomes less easy to define exactly the size of the population directly dependent

on a given land area.    Most difficult of all is the situation in
modern industrial societies.    In the UK, for example, 98% of the
population do not work in agriculture, but nevertheless the
existence of an agricultural sector in the economy does provide
employment for many others, in industries supplying goods like
fertilizers and services like government administration (or interference!).

The gross energy productivity statistic is not one that appears
widely in the literature, which usually focuses on variations in
net output per person-hour of actual labour (e.g., Clark &
Haswell 1970, Leach 1976).    It has generally been found that the
pre-industrial and semi-industrial farming systems are again not
particularly different in labour productivity.    The former ranges
from 11—40 MJ/person-hour, the latter from 23—48 MJ/person-hour.
Full-industrial systems appear to be up to two orders of magnitude more productive as a result of labour-saving technology.
Cereal production in 1968 in the UK, for example, yielded 3040
MJ/person-hour when on-farm labour only is considered.    To a
large extent, however, this difference is illusory:    "This enormous gain (through industrial technology) is largely lost in two
main ways.    Much of the crop is fed to animals who convert it
with relatively low biological efficiency... the second loss occurs
in all the post-farm sectors of the food chain.    When all direct
and indirect labour in the entire food production chain is taken
into account, the productivity for the UK turns out to be at the
most 35 MJ/man-hour (Leach 1976, p.10).    Strict accounting
methods, to transform the conventional labour-productivity
statistic into a form which considers the food-production system
as a whole, will therefore place the labour productivity of the
UK agricultural system of 1968 well within the range that Leach
portrays for pre-industrial societies.

## 6.    SURPLUS ENERGY INCOME

Neither yields nor productivity give any necessary indication
of the standard of living of the agricultural workforce.    This is
conventionally portrayed in terms of monetary income, but if we
wish to compare societies widely separated in space and time,
some indication of the purchasing power of this income is required.

The approach adopted in this paper is to calculate the 'surplus
energy income' per person.    'Surplus' is defined as that portion
of a person's income that is not required for purposes of subsistence.    This usage avoids the difficulty that in some societies
much effort is directed towards subsistence production, so that
cash incomes appear small but are in fact all 'surplus', whereas

in other societies the cash income appears large but almost all is needed for basic food purchases.

This 'surplus' income is then converted from local monetary units into a food energy equivalent by considering its buying power in obtaining foods from local sources. These foods are those usually purchased, using local prices and preferences. The surplus energy income is averaged out over the population of workers and their dependents (like any other per capita income statistic) and will be expressed here in MJ food energy per person-day.

## 7. EXOGENOUS ENERGY RATIO

If the surplus energy income portrays the efficiency of an agricultural system from the grass-roots point of view, then what of the external viewpoint? What is of interest to the outside world is the amount of food energy that a system produces and exports, in relation to the quanitity of energy supplied, by the outside world, that it consumes. The ratio of exported energy to imported energy can be termed the 'exogenous energy ratio' $(E_{ex})$.

In an earlier study, this ratio was calculated for village populations in Fiji and the Solomon Islands, where one major export crop, copra, was being produced at a time of retail price inflation and copra price fluctuation in the early 1970's (Bayliss-Smith 1977). It was found that in these part-subsistence, part-market economies, the combination of falling copra prices and inflation caused a reduction in exported output, but at the same time caused a much larger reduction in the consumption of external products, such as food and fuel. The net effect on Pacific islanders of these price changes (typical of those current in the agrarian Third World) was therefore to increase their efficiency in terms of the exogenous energy ratio, even though their own living standards were suffering.

## 8. TOTAL ENERGY RATIO

Perhaps less revealing, because more generalized, is the 'total energy ratio' $(E_r)$ first discussed in a comparative way by Harris (1971), Lawton (1973) and Slesser (1973), and illustrated for some 80 different agricultural systems by Leach (1976). This ratio is the total net food energy output divided by the total anthropogenic input of energy, predominantly labour for pre-

industrial systems but increasingly dominated by the various
subsidies as industrialization proceeds.    Leach's findings are
summarized in figure 2, which shows the range of total outputs
and total inputs encountered, and hence the range of efficiency
ratios.    The farming systems are divided into pre-industrial,
semi-industrial and full-industrial, this last category being sub-
divided into arable systems and mixed systems (i.e., with animals
included).

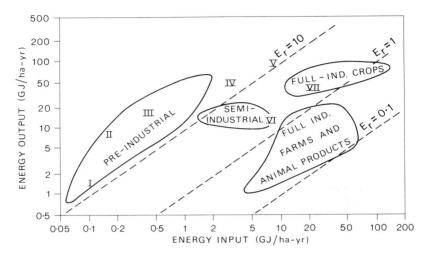

FIGURE 2.    Summary of Leach's (1976) data on energy inputs
(work and subsidies) and net outputs of food energy
from 84 systems.    I-VII show case studies of
present paper.

We can summarize the total energy ratios that are protrayed in
figure 2 as follows.

| | |
|---|---|
| Pre-industrial crops: | $E_r$ = 13—65 |
| Semi-industrial crops: | $E_r$ = 1.7—11 |
| Full-industrial arable: | $E_r$ = 0.7—4.9 |
| Full-industrial/mixed animal systems: | $E_r$ = 0.1—1.2 |

In energy terms, clearly the Industrial Revolution has begun
to encounter severely diminishing returns.    In agriculture, the
substitution of energy subsidies for human labour has increased

inputs much more substantially than it has outputs.    For example, 60 years ago in the UK, the initial electrification of farms re-quired an energy input of about 10—20 MJ per labour hour saved. Today, the equivalent figure for all UK farming is about 230 MJ extra input per additional man-hour saved.    Meanwhile, UK food energy yields have risen only about 50% during this period. UK agriculture is now comparable to heavy engineering in terms of energy use per worker (Leach 1976).

## 9.   SEVEN CASE STUDIES

In very few cases do we have sufficient data from one system for all five of these indices of efficiency to be calculated. Numerous agronomic studies provide us with details of inputs and outputs of different crops, but little data on actual land-use patterns or about the agrarian community that manages the agri-cultural system.    Similarly, geographical studies of land use are usually weak on the functioning of the agricultural systems or their social aspects.    Sociological and anthropological studies seldom provide quantitative data on land use or agronomy.

The seven case studies that have been chosen are sufficiently complete for all the relevant aspects of the agricultural system to be quantified.    They can also be taken as being representa-tive of six of the major types of social organization of agriculture (figure 3).    In more isolated parts of the pre-industrial world, we have examples of small-scale societies, virtually self-sufficient with production entirely for exchange (e.g., New Guinea).   An enlargement in the scale of agrarian society leads to some pene-tration of market influences, and encourages the development of social stratification.    Instead of being based on reciprocal ex-change, the relations between the farmer and those who help him become unequal, eventually being codified in caste distinctions of patron and client (e.g., South India, circa 1955).    The influence of modern capitalism has impinged rather differently in these two sorts of society.    Small-scale subsistence societies have, in the face of Western contact, acquired a dual character, with externally orientated cash sectors coexisting with the traditional agrarian society (e.g., Ontong Java Atoll, Solomon Islands).    Further 'modernization' of such societies leads to changes similar to those accomplished by the Green Revolution in more 'feudal' societies:   the creation of a market-dominated society of peasants having capitalist-type relations to their workforce, who are predominantly landless labourers (e.g., South India, circa 1975).

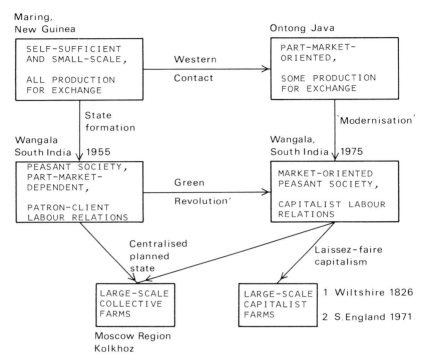

FIGURE 3.    Six major categories in the social organization of
agriculture, and the seven case studies which
represent them.

The Industrial Revolution and accompanying social changes
has led to the emergence of two divergent types of modern farming system.   In the capitalist world, engrossment of peasant
holdings led to large-scale capitalist farms, initially with a
largely pre-industrial technology (e.g., Wiltshire, southern
England, circa 1826).    In recent decades such farms have had
their manpower and workpower increasingly supplemented and
replaced by industrial technology (e.g., Southern England,
circa 1971).    Other peasant societies were propelled into a very
different mode of production, through collectivization imposed
by communist regimes (e.g., Moscow Region Kolkhoz).
    If the indices of efficiency that have been outlined do not

enable us to discriminate between these different types, one of
two conclusions could be drawn.   We could either conclude that,
despite radically different methods of organizing production,
fundamental similarities in performance have been revealed.
Alternatively, we could reject our methods of evaluation as being
of little use.    Discussion of which response is the more reasonable must await presentation of the data.

## 9.1.    Land use, population and inputs

   Further details of these seven case studies are shown in table
1.    In Leach's terms, two are pre-industrial (the Maring of New
Guinea and Wiltshire in 1826), three are semi-industrial (Ontong
Java Atoll, and South India pre-and post-Green Revolution), and
two are full-industrial systems (Moscow Region collective, and
southern England in 1971).    The percentage of total energy
input derived from fossil fuels thus varies from 0 to 99.8%, as
does the proportion of output sold to external markets.    The
population and area exploited within each farming system vary
considerably, but all have two things in common:   they are
integrated systems, all defined in the same way;   and all are
mixed farms with no particular emphasis on animal products.
   The levels of inputs per hectare and per capita during the
year to which the data refer are shown in table 2.    The energy
of human work is expressed both in hours per week per adult
and in its approximate energy equivalent (details of calculation
procedures are given in Bayliss-Smith, in press).    The workload varies remarkably, ranging from 9.5 hours per man and
woman in the New Guinea society to 50 hours for agricultural
workers in England in 1971.    The other societies are on average
intermediate, but individuals within them would be comparable
to both ends of this scale according to whether they were fully
employed or only part-time or casual labourers.
   When considered on a per hectare basis, the labour inputs
reveal a very different pattern.    In pre-industrial or Third
World societies, the main control over the input level is the
degree of population pressure on land.    This pressure is slight
in the case of New Guinea and 19th century Wiltshire (but for
very different reasons, horses and unemployment severely
reducing the number of persons supported by the English farm),
but very substantial in South India, where the wet-rice systems
soak up massive amounts of labour. Much of the smallest labour
inputs per hectare are on industrialized farms in southern England, where mechanization has diminished labour needs to only

TABLE 1.   Summary data for seven agrarian communities and
their agricultural systems.

| Name | Classification (% of total energy input from fossil fuels) | Population directly supported by agriculture | Social organiza- tion |
|------|------------------------------------------------------------|----------------------------------------------|------------------------|
| (a) Maring, New Guinea | Pre-industrial (0%) | 204 | Clan, communal land owner- ship, dispersed hamlets |
| (b) Wiltshire, 1826 | Pre-industrial (5%) | 128 | Large farm, early capitalist period |
| (c) Ontong Java | Semi-industrial (13%) | 850 | Nucleated settlements, communal land tenure by clans |
| (d) Wangala, 1955 | Semi-industrial (58%) | 7.5 | Typical farm, peasant caste |
| (e) Wangala, 1975 | Semi-industrial (78%) | 16 | as (d) |
| (f) Moscow Kolkhoz | Full-industrial (96%) | 2005 | Russian collec- tive farm (stat- istical average) |
| (g) Southern England, 1971 | Full-industrial (99.8%) | 25 | Large farm, modern capita- list (statistical average) |

Principal sources: (a) Rappaport (1968, 1971), (b) Cobbett (1830),
(c) Bayliss-Smith (1977), (d) Epstein (1962, 1973), (e) Rebello
et al. (1976), (f) Moscovskaia Oblast (1967), (g) Ministry of
Agriculture (1973), Leach (1976).

| Environment | Major outputs | |
| --- | --- | --- |
| | Subsistence | Market |
| Lower montane forest | Taro, yams, sweet potato, pigs | – |
| Chalk and vale | Bread, beer bacon | Wheat, barley, sheep |
| Atoll | Taro, coconut, fish | Copra |
| Deccan plateau | Rice, millet | Sugar cane |
| as (d) | Millet | Sugar cane, rice |
| Podzols, mixed forest | Potatoes, vegetables, milk, pork | Wheat, potatoes, cabbages, milk |
| Average, Southern England | – | Barley, wheat, cattle, sheep, milk, eggs |

TABLE 2. Energy inputs in seven agrarian systems (in MJ).

| Name | Hours worked in agriculture, per adult worker, per week | Energy input per hectare per year | | | Energy input per capita[1] per year | | |
|---|---|---|---|---|---|---|---|
| | | Human work | Imported subsidies | Total | Human work | Imported subsidies | Total |
| (a) Maring, New Guinea | Men and women, 9.5 hours | 103 | 0 | 103 | 208 | 0 | 208 |
| (b) Wiltshire, 1826 | 37 hours[2] | 140 | 43 | 183 | 504 | 154 | 658 |
| (c) Ontong Java | Men 23 hours, women 14 hours | 393 | 69 | 462 | 425 | 64 | 488 |
| (d) Wangala, 1955 | 42 hours[2] | 1379 | 1876 | 3255 | 578 | 786 | 1364 |
| (e) Wangala, 1975 | 41 hours[2] | 1610 | 5268 | 6878 | 318 | 1040 | 1358 |
| (f) Moscow Kolkhoz | 43 hours[3] | 226 | 5919 | 6145 | 355 | 9281 | 9636 |
| (g) S.England, 1971 | 50 hours | 35 | 21 835 | 21 870 | 652 | 401 760 | 402 412 |

[1] Entire population directly supported (i.e. on-farm workers and their dependents).

[3] Per household, many of which contain more than one worker.

[2] Average figure, hiding great variations between full-time and part-time workers.

one-third the level for shifting cultivators in the forests of New Guinea.

Quite soon in the development process, these inputs of work energy become dwarfed by the energy needed to produce other inputs, such as fertilizer, irrigation water and machinery. In industrialized countries, farm workers have at their disposal about 2000 times more energy than their counterparts in self-sufficient pre-industrial systems. The land itself receives some 200 times more energy input per hectare, but this is only one-tenth the increase in per-capita energy use. We can conclude that as well as having a greedy appetite for fossil-fuel energy, industrial farms also employ very few people by comparison with their pre-industrial predecessors.


## 9.2. Output and yield variations

The variation in output between the seven agricultural systems can be expressed in either gross or net form. In both cases, we are discussing the output of food energy edible by humans, but in some farming systems a proportion of the gross production of food is recycled, and much of it dissipated, through animal husbandry. In many New Guinea societies, for example, a substantial surplus of sweet potatoes is produced to maintain large pig herds. Barley and oats were (and are) grown on English farms for direct feeding to livestock. In other cases, animals manage to survive on products unfit for human consumption: cattle and buffaloes in India survive largely on waste products like sugar-cane leaves and by grazing on waste ground.

In table 3 are shown both gross and net outputs, for the seven case studies under review. In all but the New Guinea case, the exported food energy is a high proportion of the total produced, an indication of the strong dependency of all but the most remote agricultural systems upon trade with the outside world.

The food energy yield ranges from 1.5 GJ/ha in the case of the Maring in New Guinea to 66.5 GJ/ha for Wangala, South India, in 1975. For the farming systems within the tropics (a,c,d and e), position within this range of yields reflects the extent to which technology, both traditional and industrial, has been applied to solve two problems: one is the need to maintain the level of soil nutrients, and the second is a perennial water supply in order to maximize the proportion of the year that the land remains productive.

For temperate areas, the unproductive winter period imposes a lower ceiling on food energy yields, but in practice a more potent factor is the animal husbandry emphasis that typifies

TABLE 3.  Energy outputs from seven agrarian systems (in GJ).

| Name | Net output of food energy [1] | | Output exported as % of total gross output |
|---|---|---|---|
| | GJ/ha | GJ/person–year | |
| (a) Maring, New Guinea | 1.5 | 3.0 | 0% |
| (b) Wiltshire, 1826 | 7.4 | 26.6 | 98% |
| (c) Ontong Java | 14.8 | 14.0 | 86% |
| (d) Wangala, 1955 | 42.3 | 17.7 | 88% |
| (e) Wangala, 1975 | 66.5 | 13.1 | 99% |
| (f) Moscow Kolkhoz | 8.1 | 12.7 | 92% |
| (g) Southern England, 1971 | 44.9 | 883.2 | 100% |

[1] Crops and animal products consumed by the population or sold. Gross outputs, which include also potentially edible foods produced and fed to animals (not relevant for (c), (d) and (e)) are as follows: (a) 1.9GJ/ha, 3.8 GJ/person–year; (b) 8.2 GJ/ha, 29.3 GJ/person–year; (f) 13.6 GJ/ha, 21.4 GJ/person–year; (g) 48.0 GJ/ha, 886.0 GJ/person–year.

temperate farming in general.    Pre-industrial Wiltshire shows a yield of 7.4 GJ/ha, which 150 years later had increased six-fold, the change being the result of an increased proportion of arable land as much as arable yield increases as such.    By comparison, the collective farm in the Moscow Region has food energy yields of only 8.1 GJ/ha, much closer to pre-industrial than to present-day English levels.    In the Moscow case, the importance of milk and vegetable production for the urban market is part of the explanation for this poor performance, which also results from poor soils, a short growing season, and a very modest level of energy subsidy.    In 1966, the poor organization of collective farms might well have contributed to their inefficiency.


9.3.    Productivity and technological change

The food output per capita statistic, or gross energy productivity is shown in table 4.    As with the yield data, some fascinating similarities are revealed between systems which in other respects would appear to be quite different.    The two societies dependent upon manual labour, New Guinea and Ontong Java, span the approximate range 10—40 MJ output per person-day, a reflection largely of the predominance of the coconut in the total output of Ontong Java.    Coconut, exported from the atoll in dried form, is an energy-rich product requiring relatively little labour except in the harvesting and processing of the nuts.

With one exception, all the societies in table 4 are in the range 36—79 MJ/person-day, the highest productivity being in pre-industrial Wiltshire.    Here, the data fully support William Cobbett's remark, that "our indignation and rage against this infernal system is not fully roused till we see the small number of labourers who raise all this food and drink".    Cobbett recognized that the high productivity of the workers (enhanced by abundant horse power) would have enabled them to be far better fed and clothed, were it not for the "infernal system" of early capitalism.    In a contemporary peasant society exposed to the same process, Wangala village in southern India before and after the Green Revolution, productivity seems to have fallen over a 20 year period despite rising yields.    Industrialization in this case has not yet been accompanied by labour-saving technology, which is redundant in a situation of population growth and chronic underemployment. The average peasant farmer requires more labour since the introduction of high-yielding varieties of rice, but it is supplied by an expanding pool of landless peasants and untouchables, who in consequence are increasingly underemployed.

Southern England in 1971 provides the major exception to the general productivity levels discussed above.    Here, the gross

TABLE 4.    Indices of efficiency for seven agrarian systems.

| Name | Food energy yield | Gross energy productivity |
|------|-------------------|---------------------------|
|      | MJ/ha             | MJ/person-day             |
| (a) Maring, New Guinea | 1460 | 10 |
| (b) Wiltshire, 1826 | 7390 | 80 |
| (c) Ontong Java | 14780[1] | 39 |
| (d) Wangala, 1955 | 42280 | 49 |
| (e) Wangala, 1975 | 66460 | 36 |
| (f) Moscow Kolkhoz | 8060 | 59 |
| (g) Southern England, 1971 | 44890 | 2427 |

[1]Data exclude outputs from marine resources.

[2]Surplus of food rather than of money (i.e. root crops produced to maintain pig herds).

food output per person-day is in excess of 2400 MJ, a 30-fold increase on the level 150 years ago.   To some extent this increase exaggerates the labour-saving effect of industrial technology, since many food-related jobs are created indirectly by agriculture in the manufacturing and tertiary sectors.   The question never-theless arises as to how appropriate this technology is in a world where an increasing population is seeking meaningful employment as well as mere survival.


9.4.    The expropriation of productivity:   real incomes

Whereas the productivity figure represents potential income, the 'surplus energy income' statistic portrays the actual situa-

| Surplus energy income | Exogenous energy ratio | Total energy ratio |
| --- | --- | --- |
| MJ/person-day | Exported output / Imported input | Total net output / Total input |
| 2.3$^2$ (average) | – | 14.2 |
| 2.4 (labourer) | 169 | 40.4 |
| 5.3 (average) | 192 | 28.8 |
| 7.5 (farmer's household) | 22 | 13.0 |
| no data | 12.5 | 9.7 |
| 4.1 (average) | 1.2 | 1.3 |
| 18.8 (labourer) | 2.1 | 2.1 |

tion after subsistence food needs have been met.    A comparison of the two suggests the scale of expropriation of local food surpluses by society as a whole.

Not all the data on surplus in table 4 can easily be compared, since in stratified societies average incomes can be misleading. It is notable, however, that the surplus of pre-industrial New Guinea villagers differs hardly at all from that of agricultural labourers in Wiltshire in 1826 (2.3 compared to 2.4 MJ surplus income, per person-day).    This was the reason for Cobbett's polemic, sparked off by the sight of abject poverty amongst those producing the bumper harvest of 1826, because of "the mere trifling portion of that they are suffered to retain for their own use".

A higher level of surplus energy income prevails on Ontong Java atoll (5.3 MJ), after a century of partial involvement in

market trading.   The cost of this improved welfare is the increased dependency on external conditions, which in the last
decade have resulted in a steady deterioration in the terms of
trade for Third World farmers.   The peasants on collective farms
in Moscow Region would appear to earn in real terms rather less
than their contemporaries like the islanders of Ontong Java, but
their position is at least safeguarded by the heavy but paternal
hand of the communist state.   Their position is undoubtedly more
favourable than that of labourers on peasant farms in India, although data are not available to substantiate this point.

   Farm workers in the UK, although amongst the worst-paid in
the industrial economy, have at least seen a massive improvement
in incomes since the 19th century.   Their surplus energy incomes
of 18.8 MJ in 1971 are four times greater than those of their
Soviet counterparts.   At the same time, the gap between what
they produce (gross energy productivity) and what they consume
(surplus energy income) has widened even further.

## 9.5    The external view

   A further trend is illustrated by the 'exogenous energy ratio',
which shows the performance of an agricultural system in producing food for the outside world in relation to the quantity of
energy with which it is supplied.   Industrialization is seen in
terms of this ratio to involve steadily diminishing returns to
energy supplied.   Whereas in 1826, 169 units of food were exported from farms for every unit of energy supplied (mainly coal
for blacksmiths), the ratio had fallen to 2.1 in 1971.   In the
20 years 1955—75, Wangala, South India, almost halved its
exogenous efficiency as its farmers began to adopt the package
of technological innovations associated with the Green Revolution.
In the light of the oil-based energy source that underlies so
much of industrialized technology, this trend is in many ways an
alarming one.

   The 'total energy ratio' encapsulates many of these variations
in efficiency which have been discussed.   The values shown in
table 4 do not contradict the pattern described by Leach (1976)
and summarized in figure 2.   However, this paper argues that
in itself this is an inadequate index of efficiency, obscuring much
that is of significance.   The other four measures that have been
described and illustrated above would seem to provide at least
a minimal picture.   Not surprisingly, our initial question about
the world's most efficient farmers proves to be unanswerable
except in qualified terms.

REFERENCES

Bayliss-Smith, T., 1977, Energy and Economic development in Pacific communities. In Subsistence and Survival: Rural Ecology in the Pacific, edited by T. Bayliss-Smith and R.G.A. Feachem (London: Academic Press), pp.317-359.

Bayliss-Smith, T., 1982, Ecology of Agricultural Systems (Cambridge: Cambridge University Press), in the press.

Clark, C. & Haswell, M., 1970, The Economics of Subsistence Agriculture, 4th Ed. (London: Macmillan).

Cobbett, W., 1830, Rural Rides, reprinted 1967 (London: Penguin).

Epstein, T.S., 1962, Economic Development and Social Change in South India (Manchester: Manchester University Press).

Epstein, T.S., 1973, South India: Yesterday, Today and Tomorrow (London: Macmillan).

Harris, M., 1971, Culture, Man and Nature (New York: T. Crowell).

Lawton, J.H., 1973, The energy cost of food gathering. In Resources and Population, edited by B. Benjamin, P.R. Cox and J. Peel (London: Academic Press), pp.59-76.

Leach, G., 1976, Energy and Food Production (Guildford: IPC Science and Technology Press).

Ministry of Agriculture, 1973, Farm Incomes in England and Wales, 1971-2 (London: HMSO).

Moscovskaia Oblast, 1967, Moscovskaia Oblast' Za 50 Let Statisticheskiy Sbornik (Moscow: Statistika).

Pimentel, D. & Pimentel, M., 1979, Food, Energy and Society (London: Arnold).

Rappaport, R.A., 1968, Pigs for the Ancestors: Ritual in the Ecology of a New Guinea People (New Haven: Yale University Press).

Rappaport, R.A., 1971, The flow of energy in an agricultural society. In Energy and Power (San Francisco: W.H. Freeman), pp.69-82.

Rebello, N.S.P., Chandrashekar, G.S., Shankaramurthy, H.G. & Hiremath, K.C., 1976, Impact of the increase in prices of inputs in Mandya District of Karnataka. Indian Journal of Agricultural Economics, 31(3), 71-81.

Slesser, M., 1973, Energy subsidy as a criterion in food policy planning. Journal of the Science of Food and Agriculture, 24, 1193-1207.

# THE COMPARATIVE ECONOMICS OF AGRICULTURAL SYSTEMS

C.R.W. Spedding

Department of Agriculture & Horticulture,
University of Reading,
Whiteknights, Reading RG6 2AN

## 1. INTRODUCTION

Agriculture is a purposeful activity (figure 1) but is commonly carried out for several different purposes, not aways consistent with each other. In this, of course, it is no different from many - perhaps most - human activities, but when it is primarily concerned with meeting essential needs the resolution of these inconsistencies has to be on an economic basis. That is to say, when an agricultural system is operated primarily to provide a household with food (as in subsistence farming), it must produce more resources than it uses, unless the resources used are free. Thus, solar energy is used in larger quantities than are represented in the energy harvested in the food grown, but the food energy and protein harvested must be greater than that expended by the farmer and his family in growing the crop.

Not only is this obviously necessary in order to sustain the household in both its farming and non-farming activities, but not all the energy in the food harvested is available to Man, even if all the food is consumed.

Solar energy may be regarded as a free resource, although collecting it is dependent on possessing an area of land and suitable plants, both of which cost something. Carbon dioxide, atmospheric nitrogen, oxygen and rainwater are also free resources, necessary for crop production, and are similarly dependent on a receiving surface. In some systems, such as shifting cultivation, other non-purchased resources are used, in the form of soil nutrients; but here the area of land is even greater, since a new area has to be cultivated every year or so. In fallow systems, the cultivated area is returned to after a period of seven to 30 years, by which time the fertility has been restored.

Some resources, however, are remarkable in the sense that the amounts used are not necessarily dependent upon the area of land or the rate at which the resources are used. All the

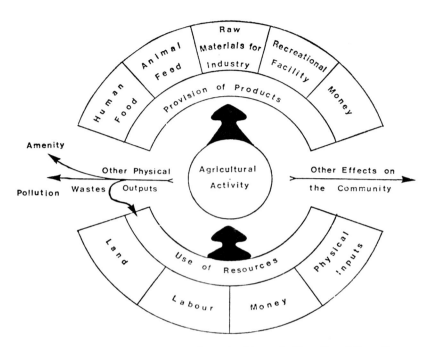

FIGURE 1.    The purposes of agriculture (after Spedding &
             Walsingham 1976).

atmospheric gases are of this kind, their replenishment being
powered indirectly by solar radiation.

   In primitive agricultural systems the major agricultural
resources, of land, labour, plants, animals, solar radiation and
atmospheric gases, are linked together.   Given that the land is
owned by the farmer, these resources can be used to provide
food, fuel, shelter and clothing, without recourse to money.
This does not mean that exchange of goods is impossible (by
barter) or that resources can be used any less efficiently.   In
most situations, however, farms do not provide for all the
different needs of the household or even for their own operation:
inputs have to be purchased and products have therefore to be
sold for cash.    Equally, because farming is part of a cash
economy it does not mean that the 'way of life' is not an important part of the total purpose.

Whatever the nature of the agricultural system, however, it will always be concerned with the efficient use of scarce resources: it is thus fundamentally involved in an economic activity, whether it uses money or not. Developed agricultural systems, of course, use vast amounts of money (table 1) and must operate like any other business, producing an adequate return on the capital invested.

TABLE 1. Money used in developed agricultural systems: examples of capital required to set up a modern enterprise.

| Enterprise | Set-up capital (£) |
|---|---|
| Dairying (per cow) | 1615 |
| Pig production (per sow) | 2430 |
| Egg production (per hen) | 9.50 |
| Barley production (per ha) | 565 |

Concern with efficiency does not, of course, imply the achievement of any particular level of efficiency.

In energetic terms, less will always be produced than is used, and wastage is inevitable. In monetary terms, it is always necessary to produce more than has been spent. Much of farming consists of converting one form of energy into another (more useful, more valuable or more desirable) and elaborating simple chemical substances into more complex ones.

The efficiency that has to be achieved must be related to the purposes of farming and to the weighting given to each of them, but it is clear that no-one will aim to produce more of something, except per unit of something else. Furthermore, there will generally be several resources involved and it will often be the case that increasing the efficiency with which one resource is used will reduce the efficiency with which others are employed. Emphasis will usually have to be given to those resources that are scarce and these will generally also be the most costly.

Sometimes, however, scarcity in the long term may not be reflected by costs in the short term. This kind of discrepancy has been the case with oil and has led to distortions (again, not only in agriculture), where agricultural systems have developed in certain directions because oil was relatively cheap.

## 2. THE EVOLUTION OF AGRICULTURAL SYSTEMS

One of the most curious distortions concerns the reduction that has taken place in the use of biological processes within agriculture (table 2). Since any definition of agriculture will include the use of plants and animals (and possibly land) for the purposes of agricultural production (and not merely of food), it is strange that the development of agricultural systems has involved this decrease in biological content. Yet even a cursory consideration of, say, battery egg production makes clear what has happened. The hen remains in its cage, with all its needs (food and water) brought to it and its eggs and faeces removed, often automatically in the latter case. The hen no longer incubates its own eggs, collects its own feed or distributes its own faeces. Indeed, anything that can be done for it is done: it only does for itself what only it can do and, even then, hormones or light control may be used to influence its physiology.

TABLE 2.   The reduction of biological processes within agriculture.

| Biological process | Replacement |
|---|---|
| Hatching eggs under a hen | Electric incubators |
| Foraging for feed by pigs and poultry | Factory feed preparation and mechanized feeding |
| Calf rearing on the cow | Artificial rearing on milk substitutes |
| Natural service by bulls | Artificial insemination |
| Weeding by hand or man-driven machine | Application of herbicides |
| Horse-ploughing | Ploughing by tractor |
| Fixation of nitrogen by bacteria | Manufacture and application of 'artificial' fertilizers |

This is not being presented here as right or wrong, merely to illustrate what has happened historically in the 'developed' agricultural systems.

There are, of course, moral issues involved and everyone is entitled to a view, but at this point I merely wish to recognize what has happened and why.

The reasons are of two main kinds. Firstly, there are good agricultural arguments for greater control over the main biological processes. There are enormous advantages in controlling what a hen eats and where it lays its eggs. In many of these cases, it is also in the interests of the hen, for example, in terms of adequate nutrition and disease control. The moral arguments are not really about the principle of control, but about the degree of interference with natural behaviour that is permissible.

Secondly, greater profitability results from greater control and from the substitution of machines and fuel for labour (figure 2). All this has been made possible by the relative abundance and low cost of oil, a situation now rapidly changing.

Currently, however, the agricultural systems of the developed world depend to a considerable extent on fossil fuels to make machines and to power them, and to manufacture all the other inputs now required to sustain current levels of productivity per unit of land.

So the present position is characterized by large inputs of 'support' energy (table 3) relative to the energy output in the form of products. Now, it is not only energy that is being produced, of course, and the form of the energy is also important: energetic efficiencies (table 4) may not therefore be of major importance as measures of agricultural efficiency, but the support energy used per unit of product is important.

TABLE 3. Inputs of 'support energy' to agriculture.

| Production system | Support energy used (MJ/kg) | |
|---|---|---|
| | UK[1] | USA[2] |
| Milk | 4.8–7.59 | 7.3 |
| Beef | 88 | 46.3–90.0[3] |
| Egg | 50.2 | 33.22 |
| Wheat | 4.3 | 5.09 |
| Potato | 1.52 | 3.17 |

[1] From Leach (1976), Spedding & Walsingham (1975).
[2] Calculated from table 4.5 in Fluck & Baird (1980).
[3] Feedlot beef.

*Energy and Effort*

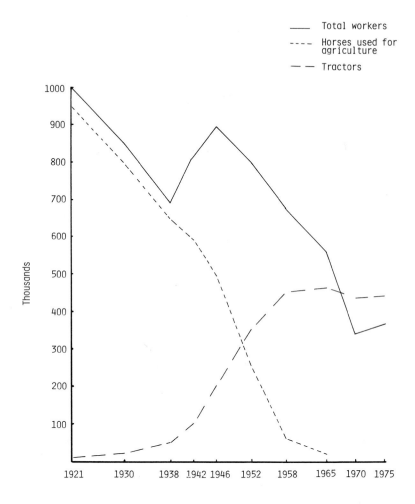

FIGURE 2.    Reduction of labour in agriculture and increase in mechanization.    Totals for 1921—1965 are from MAFF and DAFS (1968) and relate to Great Britain, totals for 1970 and 1975 are from CSO (1980) and relate to the UK.

TABLE 4.   Energetic efficiencies (E) in agriculture (E = gross energy in food produced/support energy used to produce it).

| Product | E |
|---------|---|
| Milk | 0.33—0.62 |
| Beef | c. 0.18 |
| Broilers | c. 0.1 |
| Eggs | c. 0.16 |
| Wheat | 2.2—4.6 |
| Barley | c. 1.8 |
| Potatoes | 1.0—3.5 |

TABLE 5.   Examples of possible 'new' biological processes for agriculture.

| Purpose | Biological process |
|---------|--------------------|
| Recycling of wastes | Culture of earthworms*, dipterans* or fungi |
| Conversion of crop by-products | Culture of snails* |
| Production of feed for poultry | Insect culture |
| Drying manures (slurry, etc.) | Cultures of invertebrates* |
| Production of food or feed for animals from fibrous plant materials | Enzyme or microbiological treatment of cellulose |
| Food or feed production | Controlled growth of algae in water cultures |

* Possible feeds for pigs, poultry or fish.

*Energy and Effort*

Now that oil is becoming both scarce and expensive, agriculturalists will have to consider how this dependence on oil can be reduced and, while there may be no case for going back to old methods, particularly those that involve human drudgery, part of the solution may involve the skilful harnessing of <u>more</u> biological processes. These might include developments of past methods but also quite new ones (table 5).

## 3. THE USE OF ENERGY IN AGRICULTURAL SYSTEMS

In addition to the support-energy inputs already mentioned, agriculture is the major user (other than the community at large) of solar radiation (table 6), which appears to be used very inefficiently. However, this is a good illustration of the dangers of looking at one (partial) user of a resource as if the whole resource could be regarded as available to it (in this case agriculture). Not only is about half of the received solar radiation of the wrong wavelength for photosynthesis, but we all depend for our survival upon the light and heat that is <u>not</u> used by plants. Even that which is used by plants in, for example, transpiration, is actually contributing to the energetics of the hydrological cycle that ensures our water supply.

TABLE 6. The use of solar radiation in agriculture (from Duckham et al. 1976).

|  | MJ/annum |
|---|---|
| Energy derived* from solar radiation used in world agriculture (as cerals, roots, pulses, nuts, fruit, vegetables and their oils, sugar, meat and milk) | $29 \times 10^{15}$ |
| Support energy used (world total) | $204 \times 10^{15}$ |

* This is the energy contained in the products: the energy actually used to produce them would be very much larger.

This does not mean that we are not interested in the efficiency with which crop plants use the photosynthetically active part of

the solar radiation received by agricultural land: on the
contrary, it remains the major energy source for agriculture and
it is important that, not only is it used efficiently, but that <u>more</u>
of what reaches agricultural land is used (this is, of course,
the same thing if efficiency is calculated on the total amount
received by the land annually).

One of the most important ways of raising this efficiency is by
increasing crop yield per unit area of land (table 7). This is
because the solar radiation input remains the same, whatever
the output. Efficiency therefore rises with output. As it
happens, for many crops the most important way of increasing
yields is by the application of fertilizer, itself a very costly
input to manufacture, especially in energetic terms.

TABLE 7. The effect of crop yield on the efficiency (E) with
which solar radiation is used (E = energy in crop
per ha/solar radiation falling on one ha[1]).

| Crop yield[2] (kg DM/ha)/y | E |
|---|---|
| 2783 | 0.0016 |
| 4180 | 0.0023 |
| 6402 | 0.0036 |
| 9944 | 0.0056 |
| 11 616 | 0.0065 |
| 13 134 | 0.0072 |

[1] Assumed to be $33 \times 10^6$ MJ/ha/y.
[2] With different levels of nitrogenous fertilizer (0-400
kg/ha/y).

Thus, it turns out that one powerful way of increasing the
efficiency with which solar radiation is used is by inputs of
additional, support energy (figure 3). In this way, overall
energetic efficiency may be improved but the efficiency with
which fossil fuels are used may fall.

This illustrates two important points. One is that all units
of energy are not equal and the value and importance of an
energy source varies with the nature of the source. The second
point is that only in monetary terms can we currently express
these different values.

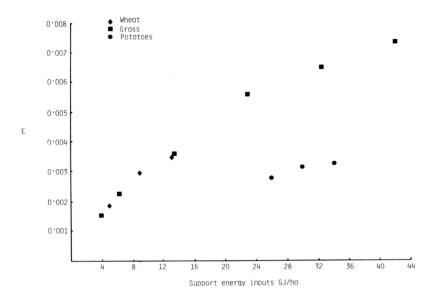

FIGURE 3.    Effect of support energy inputs on efficiency (E)
            of use of solar radiation   (E = Output of energy in
            crop/incident solar radiation).

   Liquid fuels, for example, have properties that make them
particularly valuable for certain purposes (such as transport)
and in the form of fossil fuels they are effectively non-renewable
and will therefore become increasingly scarce and costly.    It is
essential to minimize dependence on such fuels and to maximize
(or at least aim at very high) efficiency in their use.    By contrast, solar energy may be regarded as a renewable (or constant)
supply that is relatively cheap.    The emphasis therefore has to
be on  using as much as possible, even if it is not used with
particularly high efficiency.
   One way of combining these objectives is to use more human
labour (or animal power) to increase or maintain output per unit
of land without the input of large quantities of fossil fuel.

## 4. LABOUR, EFFORT AND POWER

Animal power, while still of immense importance in the developing countries, seems unlikely to regain its former place in the agricultural systems of developed countries.

In other parts of the world, animal power is vital for cultivation, harvesting and transport for goods and people, both directly and by pulling carts. In many cases, the animals live very cheaply, on by-products of arable farming or on common and waste land, and themselves produce valuable by-products (milk, meat, hides, faeces for fertilizer, fuel or building materials). There is still scope for improvement in the genetics nutrition, management and even harnessing of such animals, as well as in the design of the equipment they use.

In developed agricultural systems, however, the adjustments needed to reintroduce draught animals would be considerable. It is always easier to change in the other direction, since there are far fewer difficulties and costs in disposing of unwanted animals than there are in actually creating or producing them. Usually, the initial number is very small, and breeding a large number from these takes a great deal of time (so it cannot be done quickly) and, at first, most of the animals have to be retained to increase the breeding stock. (Thus, a trend towards increasing one kind of animal often leads to an initial shortage,, and a decision to dispose of them leads initially to a plentiful supply but to a subsequent shortage).

Even if the necessary animals could be produced, there is still the cost of feeding them. Energetically, of course, feeds are derived, indirectly, from solar radiation and not from fossil fuels – depending somewhat on how these feeds are produced. But their production usually involves a diversion of solar radiation from production to animal feeding, of the order of 10-20%.

There may be agricultural systems in which this would be feasible, even for the developed countries. They would probably be characterized by availability of by-products for feeding animals, or access to non-arable land. The former might still be less efficient than the direct use of these same by-products for the production of fuel, but this would depend on a great many variables.

Animal power has to be related to farm size, since the power required to plough, for example, may require the same size of animal, independent of the area to be ploughed: this is also true of tractors, since they are not available over an infinite range of size (especially downwards, i.e., suitable for extremely small areas).

Human power, extended by suitable tools, can cope with a very wide range of tasks and areas and has probably not been much considered in this sense – of making the maximum use of human labour by assisting it with tools but without substituting fuel for human energy.   In terms of cheap oil, it was much simpler to make the complete substitution, where possible, of machines for men.

Small-scale farming, because of the increased advantages of flexibility in power and operation, may be able to make more and better use of human labour than can large farms.

Indeed, the possibilities for fuel economy and fuel production may seem rather greater on small farms than on large ones, but this may only reflect the amount of thought so far devoted to the problems.

Where small-scale farming is visualized on a part-time basis, it is even more difficult to make the appropriate economic assessments.   Part-time labour may be priced quite differently, just as food and fuel consumed by the producer may be valued in a different way, because they save expenditure and do not have to be purchased with money that has already borne taxation.

All of this regards labour as physical effort and energy expended, but mental effort may also be involved, particularly in the management of large systems.   Such management skill may greatly influence physical and economic efficiencies, and satisfaction may be an additional, relatively unquantifiable benefit.


## 5.   COMPARATIVE ECONOMICS

Currently, however important these other aspects may be, it is only possible to assess the relative physical and financial efficiencies of agricultural systems.   This reveals major discrepancies but this is to be expected, since an increase in one sort of efficiency may well lead, inevitably, to a decrease in several other sorts.

The objectives of the manager of an agricultural system always have to include the achievement of some balance of the relevant efficiency ratios, and this can only be struck by weighting ratios in order of importance.   For most people this has to be done, in the short term, in monetary terms, with some financial index, such as profit or return on capital, as the chosen criterion of success.   In the longer term, energy accounting may be useful on the basis that the scarcity and costs of energy will dominate future economics.

To the extent that this conviction leads to changes aimed at reducing vulnerability to oil prices, however, so will it reduce the relevance of energy accounting.

If we were successful in decreasing fossil fuel inputs to negligible levels and increasing the use of solar radiation, directly and indirectly, the cost of oil would diminish in importance.

Whether increased use of human labour would influence this argument would depend on, among other things, whether the cost of such labour reflected a cost of living dominated by oil prices or some quite different influences. Almost certainly, however, the cost of human labour will not reflect the amount of energy expended.

## REFERENCES

CSO, 1980, Annual Abstract of Statistics (London: HMSO).

Duckham, A.N., Jones, J.G.W. & Roberts, E.H., 1976, Food Production and Consumption, edited by Duckham, Jones and Roberts (North Holland Pub. Co.).

Fluck, R.C. & Baird, C.D., 1980, Agricultural Energetics (Connecticut, USA: AVI Pub. Co.).

Leach, G., 1976, Energy and Food Production (Guildford: IPC Science and Technology Press).

MAFF & DAFS, 1968, A Century of Agricultural Statistics Great Britain 1866–1966 (London: HMSO).

Spedding, C.R.W. & Walsingham, J.M., 1975, Energy use in agricultural systems. Span, 18, 7–9.

Spedding, C.R.W. & Walsingham, J.M., 1976, The production and use of energy. Journal of Agricultural Economics, XXVII, 19–30.